The Book of the
RELIANT

A Practical Handbook covering all Three-Wheeled models up to 1970

John Thorpe

First published 1965
Reprinted 1968
Reprinted with minor amendments 1970

ANNOUNCEMENT

By special arrangement with the original publishers of this book, Sir Isaac Pitman & Son, Ltd., of London, England, we have secured the exclusive publishing rights for this book, as well as all others in THE MOTORCYCLIST'S LIBRARY.

Included in THE MOTORCYCLIST'S LIBRARY are complete instruction manuals covering the care and operation of respective motorcycles and engines; valuable data on speed tuning, and thrilling accounts of motorcycle race events. See listing of available titles elsewhere in this edition.

We consider it a privilege to be able to offer so many fine titles to our customers.

FLOYD CLYMER
Publisher of Books Pertaining to Automobiles and Motorcycles
2125 W. PICO ST.   LOS ANGELES 6, CALIF.

INTRODUCTION

Welcome to the world of digital publishing ~ the book you now hold in your hand, while unchanged from the original edition, was printed using the latest state of the art digital technology. The advent of print-on-demand has forever changed the publishing process, never has information been so accessible and it is our hope that this book serves your informational needs for years to come. If this is your first exposure to digital publishing, we hope that you are pleased with the results. Many more titles of interest to the classic automobile and motorcycle enthusiast, collector and restorer are available via our website at www.VelocePress.com. We hope that you find this title as interesting as we do.

NOTE FROM THE PUBLISHER

The information presented is true and complete to the best of our knowledge. All recommendations are made without any guarantees on the part of the author or the publisher, who also disclaim all liability incurred with the use of this information.

TRADEMARKS

We recognize that some words, model names and designations, for example, mentioned herein are the property of the trademark holder. We use them for identification purposes only. This is not an official publication.

INFORMATION ON THE USE OF THIS PUBLICATION

This manual is an invaluable resource and a 'must have' for owners interested in performing their own maintenance. However, in today's information age we are constantly subject to changes in common practice, new technology, availability of improved materials and increased awareness of chemical toxicity. As such, it is advised that the user consult with an experienced professional prior to undertaking any procedure described herein. While every care has been taken to ensure correctness of information, it is obviously not possible to guarantee complete freedom from errors or omissions or to accept liability arising from such errors or omissions. Therefore, any individual that uses the information contained within, or elects to perform or participate in do-it-yourself repairs or modifications acknowledges that there is a risk factor involved and that the publisher or its associates cannot be held responsible for personal injury or property damage resulting from the use of the information or the outcome of such procedures.

WARNING!

One final word of advice, this publication is intended to be used as a reference guide, and when in doubt the reader should consult with a qualified technician.

Preface

DRIVERS of Reliant three-wheelers can be folk who have never held—and perhaps never will hold—a car licence. Thus the Reliant forms a stage intermediate between the inadequate bubblecar (or the uncomfortable motor-cycle combination) and the expensive car. It does in fact represent the cheapest form of true family transport—cheap to buy, cheap to tax, and cheap to run. And under current licensing rules it can be driven by car drivers and motor-cyclists alike.

Though the combination of a man-sized engine with light body and chassis gives a satisfying performance, allied with a considerable degree of comfort, the Reliant is primarily a utility vehicle. And many owners like to maintain their Reliants themselves, thus keeping their costs even lower. To help them do so is the object of this book.

There is a lot that the average man with average facilities can tackle in the way of mechanical work. Even so, not everything is within the compass of a necessarily sparsely-equipped home workshop. Consequently, where in my judgement it is better to leave well alone I have not hesitated to say so. The object of doing mechanical work oneself is normally to save time and/or money. Where there would appear to be no significant economy in either my advice has always been to let the professionals do it.

One normally finds that if sufficient attention is paid to routine maintenance and to immediate repair of minor faults, major work does not have to be done. Really serious damage or excessive wear can be traced, in perhaps nine cases out of ten, to neglect of the basic maintenance which every car needs.

I have endeavoured, therefore, to incorporate not only sufficient data to enable the private owner of a Reliant to decarbonize his power unit or renew his clutch, but also a considerable amount of advice on looking after the car. And for the help which I have had from the manufacturers —the Reliant Motor Company, Ltd.—in supplying the data from which I have worked, and the illustrations used, I make grateful acknowledgement.

JOHN THORPE

Contents

1 THE RELIANT 1
2 BASIC PRINCIPLES 3
3 THE TOOL KIT 15
4 FAULT TRACING 18
5 ROUTINE MAINTENANCE 28
6 OVERHAULING THE ENGINE 42
7 THE FUEL SYSTEM AND CARBURETTOR 66
8 WORK ON THE ELECTRICS 76
9 CLUTCH, GEARBOX, AND TRANSMISSION 89
10 STEERING AND SUSPENSION 95
11 LOOKING AFTER THE BRAKES 100
12 CHASSIS, BODY, AND TYRES 106
Appendix—FACTS AND FIGURES 112
Index 117

1 The Reliant

In concept the three-wheeler is a particularly British institution. Other nations make bubble-cars, but only Britain can now boast of a really powerful three-wheeler—the Reliant.

That the three-wheeler has had a particular appeal in this country is a fact of motoring history. The very first all-British "car" was a three-wheeler—a "trike" designed by Edward Butler back in the 1880s. By the early years of this century a Welsh firm was producing three-wheeled cars, and from around 1910 right up to the outbreak of the Second World War the three-wheeler enjoyed a very considerable popularity.

After the War, however, the combined effects of shortages and the increasing difficulty of producing a semi-luxury family car within the legal weight limit caused more and more firms to drop out of this market. Then, just when it seemed that the three-wheeler—at least in its most powerful form—would die, up came a new name. Reliant. True, this was a firm with a pre-war history of three-wheeler production. But those had all been commercial vans powered, initially, by Austin 7 engines. In 1939 Reliants introduced a version with their own power unit, and it was this engine which was to provide the mechanical basis for the shapely Regal Mark 1 which made its debut at the Earls Court Motor Show in 1953.

In concept, this was an entirely new car. Its front suspension system was advanced for its time, and even at that early stage Reliant were using a few glass-fibre panels in the bodywork. A Mark 2 version soon followed. Then, in late 1956, there came the Regal Mark 3 which was the first quantity-produced car in Europe to have a body made entirely of glass-fibre. This durable material, allied to a sturdy box-section chassis, makes the Reliant one of the longest-lasting small cars on the market—and one of the safest, too.

Although handicapped, as always, by the need to build within a weight limit Reliant continued the search for greater luxury, and by ingenious design contrived to include wind-down windows in the Regal Mark 4 which was in production until 1961, when the Regal Mark 5 took over. This model offered, amongst other refinements, improved rear-seat headroom. And finally there came the last of the line to be powered by the well-proved side-valve engine—the Regal Mark 6, with a lengthened roof line to improve still further the passenger compartment space.

If it seemed that the limit of development had been reached, then somebody had underestimated the resourcefulness of this specialist company. The next step in the battle against weight was to produce a more powerful engine which, at the same time, was smaller and lighter. The result of this design study was the ingenious overhead-valve power unit of the Reliant 3/25—the first modern under-1000 c.c. all-alloy water-cooled four-cylinder car engine. Married to a development of the Reliant chassis and to a brand-new full four-seater body with raked-forward rear window this has produced a three-wheeler which is quite capable of holding its own with four-wheeled cars with engines fifty per cent bigger, while saving on fuel, tyre, and taxation costs.

A logical development was the introduction of the Regal 3/25 Super—mechanically identical, but with improved bodywork and modified interior. And, subsequently, the raising of the engine capacity to 700 c.c. produced the even peppier Regal 3/30.

Obviously, to achieve all this something has to be sacrificed, and it is inevitable that the Reliant is not quite as roomy nor as easily accessible for maintenance as are some conventional cars. On the other hand, there is no wasted space and in some respects—notably in the comparative ease of removing the engine unit for overhaul—the Reliant scores heavily over its four-wheeled competitors. And three-wheelers of the Reliant type can be driven by holders of either car or motor-cycle licences, This fact, coupled with the very real taxation advantages enjoyed by the three-wheeler, should ensure its popularity for many years to come.

2 Basic Principles

ALTHOUGH modern vehicles are well designed, to obtain the best from them it is essential to know exactly how they work and why. This applies not only to maintenance, important though that is, but also to actual use on the road.

So far as engines are concerned, all Reliants use four-stroke units: side-valves on pre-1962 models, overhead-valves on post-1962 models. The term *four-stroke* refers to the number of working strokes in one complete cycle of operation of one cylinder of the engine. Thus in the four-stroke a working cycle consists of four strokes—in other words, the piston travels from its uppermost position to its lowest, and vice-versa, four times.

Before considering just how the engine works, it is necessary to know the names of the major components. First there is the *block* containing the *cylinders*. Each of these is, as its name implies, simply a metal cylinder. It is closed at one end by a *cylinder head* which, in the case of an overhead-valve four-stroke engine, is equipped with channels (*ports*) through which gas can flow. For each cylinder there is an *inlet port* and an *exhaust port*. These are closed by *valves*, which are held in the closed position by *valve springs*. To open the valves, a form of mechanical seesaw called a *rocker* is used—one to each valve. In the side-valve engine, however, both valves and ports are located in the cylinder block, not the head, and there are no rockers.

The lower half of the engine consists of a case, known as the *crankcase*, on which the cylinders are mounted. Carried on *bearings* inside the crankcase is the *crankshaft*. This consists of *mainshafts*, *webs*, and for each cylinder a *crankpin*. Though the mainshafts are mounted centrally, each crankpin is offset so that when the crankshaft is revolved the crankpin moves on a circular path. Thus if the crankpin is at the top of the case and the crankshaft is revolved it will not simply rotate, as will the main shafts. Instead it will at first move downwards and outwards. Once the shaft has been rotated through ninety degrees the pin will begin to move inwards while still moving downwards. After half a turn it will begin to move upwards and outwards until, in the last quarter-turn, it moves upwards and inwards.

Thus if a *connecting rod* is attached to a crankpin the end so fixed will move with the pin in just such a manner. This part of the connecting rod is usually called the *big end*—for the very obvious reason that that particular end of the rod is invariably the bigger end.

The other end of the rod also has a pin. This is the *gudgeon pin*, and it carries a *piston*, made of light alloy. This piston fits closely in the

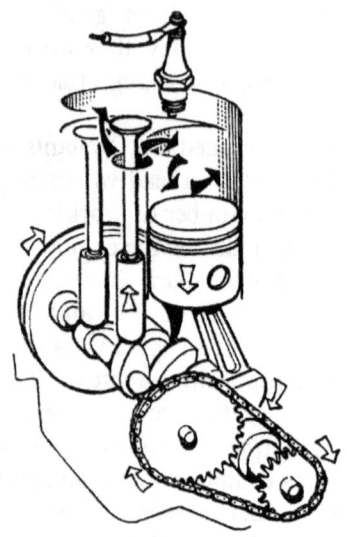

FIG. 1. THE BASIC CAR ENGINE

Movement of the piston causes the connecting rod to turn a crankshaft and so drive the flywheel. The cams controlling the valves are driven at half the speed of the crankshaft by a pair of sprockets and a chain.

cylinder, in which it is free to slide up and down. Split cast-iron *piston rings* fixed in *grooves* on the piston ensure a gas-tight seal.

If, in this basic engine, the piston is at the top of its travel it is said to be at *top dead centre*—a term usually abbreviated to t.d.c. If it is right at the bottom of its travel it is at *bottom dead centre*, or b.d.c. The distance it must travel between these two points is called the *stroke*, and this is normally measured in millimetres.

To understand the four-stroke cycle let's deal with one cylinder only, for a multi-cylinder engine is only a series of single-cylinder units coupled together. Imagine that the piston is now at t.d.c. and that the crankshaft is revolving. As it does so the crankpin moves at first downwards and outwards. This means that the big end of the connecting rod must also move downwards and outwards. Since the connecting rod cannot stretch it exerts a pull on the gudgeon pin and this in turn pulls the piston downwards. The piston being, of course, a close fit in the bore it is unable to move forwards or backwards. It can only travel up and down.

As the piston moves down the cylinder the valve which had been closing the inlet port is opened. Inside the cylinder the movement of the piston has lowered the pressure so that it is below that of the air outside, and therefore air starts to flow through the inlet port into the cylinder. On its way the air was mixed with petrol to form a combustible vapour (one which can be burned).

This induction of combustible gas (air and petrol vapour) continues for the whole period during which the piston is travelling down the cylinder, and this stroke is consequently called the *induction stroke*.

After half a revolution of the crankshaft the downward movement of the piston ends, since the crankpin must now begin to press the connecting rod upwards. Obviously if the inlet port were to be left open all the mixture which had been induced would simply be blown out again. The valve is, therefore, so arranged that it closes at this point. The rising piston now compresses the mixture and this gives the second stroke of the cycle—the *compression stroke*.

By the time the piston reaches T.D.C. on this stroke, the mixture in the cylinder has been squeezed into the tiny *combustion chamber* formed between the top of the piston—the *crown*—and the inside surface of the cylinder head. On a Reliant, this chamber will have a volume only about one-seventh that of the cylinder itself. The ratio between this and the *swept volume*—the amount of mixture induced into the engine—is called the *compression ratio*, and this is one of the vital factors in deciding the characteristics of an engine. In the case quoted here the compression ratio would be seven to one. If the mixture had been compressed into one-tenth of its original volume the ratio would have been ten to one.

Once the gas has been compressed it is ready to be burned. A spark is made and this ignites the mixture which burns rapidly. In doing so it expands, so that it can no longer be contained within the tiny combustion chamber. It exerts pressure upon every surface around it, but only one of these can move. This is the piston crown, and the effect of igniting the mixture is to create a pressure inside the combustion chamber which thrusts the piston down the cylinder on the third of its four strokes—the *power stroke*. This time there is no question of the piston being *pulled* down by the crankpin. On the power stroke it is the piston which thrusts the connecting rod down. And the rod in turn causes the crankpin to revolve, thereby turning the flywheels and rotating the main shafts. These in turn drive the vehicle through the medium of *gears* and *shafts*.

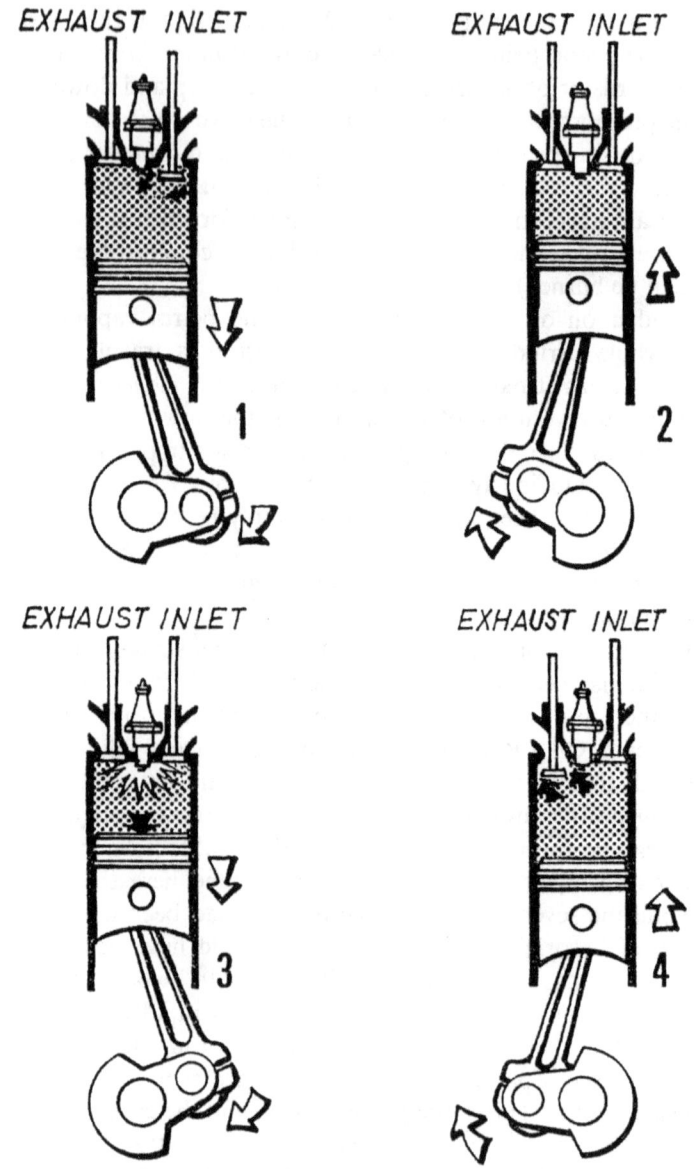

FIG. 2. THE FOUR-STROKE CYCLE
All Reliant engines, both side-valve and overhead-valve, work on this same principle
1. Induction   2. Compression   3. Power   4. Exhaust

One further stroke remains to complete the cycle—the exhaust stroke. When the piston reaches B.D.C. the driving force behind it is largely spent, and it now remains to clear the burned gases out of the cylinder. Carried by the momentum stored in the *flywheel*, the piston starts to rise. As it does so the second valve in the cylinder head—the *exhaust valve*—is opened. The rising piston pushes the burned gases up the cylinder and out of the exhaust port. At about T.D.C. the exhaust valve again closes the port, the inlet valve opens, the piston begins to descend once more, fresh mixture is induced—and another cycle of operations has begun.

That gives us the basic four-stroke cycle—*Induction, Compression, Power, Exhaust*. *Induction* and *Power* are downward strokes; *Compression* and *Exhaust* upward strokes.

A four-cylinder engine is so arranged that a different stroke is occurring in each of its four cylinders at any moment, and there are thus four power strokes to each revolution of the crankshaft. This gives a smooth-running engine.

THE VALVES

The valves of a four-stroke have to be operated mechanically, and the usual layout is to arrange for a *shaft* bearing a pair of *cams* for each cylinder to be driven, at half the engine speed, through a *chain* or by a *train of gears*. The cams bear on barrel-shaped metal *cam followers* or *tappet blocks*, and these in turn press either against an overhead-valve unit's steel or alloy rods, known as push-rods, or direct on valve tappets in the case of side-valves. Push-rods bear against one end of the *valve rockers*, and the other end of each pivoted rocker is positioned just above the *stem* of the valve. When the cam revolves it pushes the follower and rod upwards. This causes one end of the rocker to rise, and the section above the valve stem (the *tappet*) to fall. This presses the valve downwards against the pressure of its spring, and thereby opens the port. When the cam permits the rocker to return to its normal position the valve spring returns the valve to its seat, thus closing the port tightly. Side-valve operation is identical, save that the valves are opened direct, and not through the medium of rockers.

With a four-stroke, it is necessary to use an independent oiling system, fed by a pump, and this too is usually driven by the camshaft. It delivers oil from a sump through passageways to bearings and to the cylinder walls and the valve gear. The oil returns by gravity.

THE CARBURETTOR

We noted in passing that when air is induced into the cylinder it is mixed with petrol to form a combustible vapour. This, of course, is a drastic understatement of the magnitude of the job performed by a simple but precision-engineered instrument known as the *carburettor*.

In principle this is little more than a glorified scent-spray with a fancy name, but it has to carry out one of the most crucial of all jobs—metering out an exact and minute ration of petrol and mixing it thoroughly with air in just the right proportion to enable it to be burned efficiently.

At first sight this may not appear to be over-exacting, since the ideal ratio is around one part of petrol to fourteen parts of air. This, however, is the proportion by *weight*; the carburettor operates by *volume*, and on this basis each 100 c.c. of combustible vapour needs to contain only about

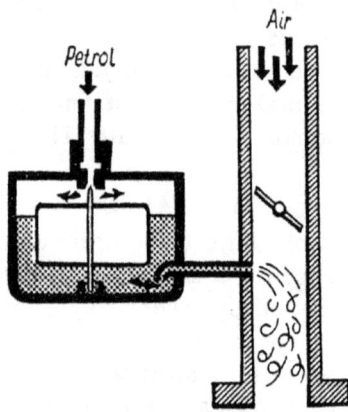

FIG. 3. HOW A CARBURETTOR WORKS
Petrol enters the float chamber through a needle-valve and enters the venturi through jets. Air is induced through the intake, the flow being controlled by a butterfly valve.

0·2 c.c. of petrol, the remaining 99·8 c.c. being air. Obviously the carburettor, despite its simplicity, is a precision instrument and has to be treated accordingly.

The basic components of a carburettor are a petrol reservoir called the *float chamber*, a *venturi* or *choke* through which air is drawn, *jets* which meter the petrol and a *throttle* which controls the amount of mixture passing through the carburettor and into the engine.

Consider the basic method of operation. Petrol is fed to the float chamber. This is very much like a pocket edition of the familiar domestic cistern. The chamber contains a *float*, which rises as petrol is admitted through a valve. In rising the float carries with it a *tapered needle*, and this needle is carefully contoured to fit in a seat in the valve. When the level of petrol in the chamber is correct the needle is pressed fully home in its seating, thus cutting off the flow of petrol. When the level in the chamber falls the float falls with it and so does the needle, thus permitting more fuel to flow into the chamber until the correct level is again reached.

Connecting the float chamber with the body of the carburettor are drilled passageways through which petrol flows. These jets, as they are called, meter petrol. Others meter air, and the two meet in the mixing chamber in the main body of the carburettor. When the induction stroke begins in the cylinder air is drawn through the carburettor venturi which is so shaped that there is a fall in pressure in the *mixing chamber*—the section around the jet tube. As a result petrol is drawn up the tube into the chamber where it mixes with the air and passes through the inlet port into the cylinder.

A carburettor which consisted of these parts alone would work, but the engine would run at only one speed unless some means of varying the supply of mixture at will was arranged. Normally, a pivotted flap called the *butterfly throttle* is employed for this, linked to the throttle pedal and spring loaded.

When the throttle is closed only a very small amount of air can pass—so small, in fact, that it is impossible for the main jet to meter out the tiny amount of fuel required. For running under these conditions a very fine jet, called the *pilot jet* delivers a minute ration of petrol to the mixing chamber.

As the throttle is opened more air passes through the venturi and jets are arranged to deliver more fuel. Since the proper action of the carburettor depends on the operation of these very fine metering devices great attention is paid to ensure that the internals are kept free from dirt. Even a speck of dirt is quite enough to block the jet and thus prevent fuel passing through it. The petrol is, therefore, normally filtered at several points by passing it through fine wire-mesh. One such filter is usually fitted in the fuel tank, a second filter in the pump, and another immediately in front of the float chamber needle valve.

When an engine is cold it needs a somewhat richer mixture than usual to enable it to start, and to supply this it is usual to employ a *choke*. This should not, of course, be confused with the venturi.

The purpose of a choke is to cut down the air supply independently of the petrol supply, thus giving the same amount of petrol but mixing it with a smaller supply of air, and a rich-mixture device of this type is invariably some form of plate which is used to block the carburettor inlet. It is a supplementary butterfly which swings over the mouth of the carburettor.

Carburettors, besides being equipped with petrol filters, also have a filter for the air. This is not so much to protect the carburettor as to protect the engine, since the air usually contains dust. And dust—harmless though it may look—comprises a surprising number of very hard particles which are quite capable of scratching the working parts of the engine very badly indeed.

An air filter itself forms an obstacle to the air flow and cuts the amount of air entering the carburettor. In the design stage this obstruction is taken into consideration and the fuel is metered accordingly. If, therefore, an instrument which is intended to have an air filter is used without one the effect is to weaken the resulting mixture, since more air is entering while the fuel supply remains unaltered. Damage to the internal parts of the engine apart, this is one reason why the engine should not be run with the air filter removed.

THE IGNITION SYSTEM

Even really experienced drivers often have only the slightest knowledge of the working of the electrical system upon which the whole operation of the engine depends. As a result the electrics are frequently neglected, failure results, and the immediate conclusion is that electricity is thoroughly unreliable anyway.

There is no need, however, to be a qualified electrical engineer to understand *how* the system works, even if the actual reasons behind it have to be taken for granted.

All electrical practice is founded upon circuits and upon the fact that an electric current will invariably take the shortest path to earth. In this connection it should be emphasized that "earth" does not necessarily mean the ground. So far as a Reliant's electrical system is concerned "earth" is the car chassis itself.

A circuit is just what its name implies, and in this electricity is rather like a model railway train. If all the points are correctly set the train will go round and round. If they are not so set it will simply end up standing still on a siding.

As with the train so with electricity. Providing there is a circuit the current will flow. If the circuit is broken it will not. And just occasionally there may be some bad points setting which directs it straight to earth—a *short circuit*—just as if the train had been directed onto a branch line leading straight to the edge of a cliff. . . .

Electricity is measured in *volts* and *amperes*. The *volt* is a measure of its force, and the *ampere* basically a measure of the number of electrons per second passing a given point. In other words while voltage indicates the electrical pressure in the circuit amperes indicate the quantity of electricity which is flowing. The resistance to the flow presented by the

BASIC PRINCIPLES

wires and so forth which make up the circuit is measured in *ohms*, one ohm being a resistance which calls for one volt to be applied so that one ampere may flow. There are also two basic types of electricity: positive and negative—but for all practical purposes it is only necessary to know that these do, in fact, exist.

Finally it is necessary to accept one further basic fact: when a coil is moved within a magnetic field an electric current is produced. There are two types of current: *alternating* and *direct*. An *a.c. generator* produces current which alternates—a constantly reversing flow. A *d.c. generator* produces current which flows in one direction only. Reliants use *coil ignition*, fed by a battery which is constantly recharged by a generator (or *dynamo*) driven by the engine.

There is one other essential part—the *distributor*. This comprises a *contact-breaker*—a mechanical switch consisting of a pair of points which are opened and closed by a cam carried on the distributor shaft—and a rotor arm and contacts. Electrically, the contact-breaker is connected into the low-tension side of the ignition circuit.

From the high-tension part of the coil a heavily-insulated high-tension lead is connected to the distributor cap and thence by a spring-loaded contact to the rotor arm. As it revolves this arm wipes contacts set in the cap, each of which is connected by cable to a *sparking plug*, set in a cylinder head. There is one plug per cylinder. This plug consists of a body, which screws into the head, and an insulated *central electrode* to which the high-tension lead is connected. Welded to the body is a *side electrode*—some plugs may have several—which is set so that a gap of around twenty-five thousandths of an inch exists between its tip and that of the central electrode.

When the ignition is switched on the battery supplies low-tension electricity to the primary winding of the ignition coil. At a pre-determined point, when the engine is turning, the cam presses one of the points of the contact-breaker away from the other and thus breaks the circuit. Here something happens which has to be taken on trust. This sudden rupturing of the low-tension circuit in the primary winding of the ignition coil creates a high-tension current in the coil's secondary winding. This current is of very high voltage—around 33,000 volts, which is the same as that carried in the conductors of the rural electricity grid system. Seeking the shortest path to earth this current streaks down the high-tension lead, through the rotor arm and contacts, and thus to the appropriate plug. Normally it would stop dead at the gap in the plug; but the pressure behind it is too great to permit it to do so. Instead it jumps across the gap in the form of a hot blue spark, and it is this spark which ignites the mixture in the cylinder. In a normal Reliant engine this operation can occur some 5000 times in every minute.

To prevent the low-tension current from doing at the contact-breaker points just what the high-tension current subsequently does at the sparking plug gap—jumping across in the form of a spark—a small electrical shock-absorber called a condenser is added to the circuit.

THE TRANSMISSION

Internal-combustion motors are high-speed engines in which power output is, within limits, proportional to the speed of rotation of the engine. At low speeds, therefore, less power is developed than at high speeds. Where outside factors, such as a hill, increase the load on an engine its speed—and consequently its power—falls off. This in turn reduces its speed still further, causing more drop in power. At length the load becomes so great that it overcomes the remaining power of the engine and the motor stalls. Basically, there is a comparatively narrow range of engine speed at which the greatest power is developed and the engine should, ideally, run at this speed wherever possible. The designer does, in fact, try to arrange for this maximum power to occur at cruising speed.

To deal with varying loads some means of keeping the engine speed up when the road speed falls is necessary, and this need is met by the *gearbox*. This consists basically of an *input* and *output shaft*, on which are carried a series of meshing gears. Each pair of gears gives a different reduction between the speeds of the two shafts. Only one pair of gears can be used to transmit the power at any one time, and in the Reliant a choice of four different gear ratios (plus reverse) is provided. Initially the drive is transmitted direct from the crankshaft to the gearbox without reduction in speed. This is reduced in the gearbox itself, depending upon which pair of gears is locked into position on the shafts, and there is a further reduction between the gearbox and the rear wheels in the *secondary transmission*. In top gear the engine crank will revolve, say, four times for each revolution of the rear wheels, but in bottom gear it may turn over twelve times. Thus one revolution of the rear wheels in top allows the force of two power strokes to be applied, but in bottom gear, in the same distance covered, the power of six strokes is passed to the driving wheels. Thus an increase in load can be counterbalanced by changing into a lower gear and bringing more power to bear in a given distance at the cost of a drop in road speed.

The method employed to lock the various gears to the shafts is simple. Certain gears are free to revolve on the shafts, and others revolve with the shaft, but are free to slide sideways on *splines*. Abutments on the side faces of the gears enable adjacent gears to be coupled together when a sliding gear is moved sideways by a *selector fork*. Selector fork movement, in turn, is controlled by the gear lever. To smooth the changes, on later gearboxes, cones on the pinions pre-engage and bring both to the same speed. This is termed *"synchromesh."*

A vital part of the transmission is the *clutch* which enables the drive to be freed at will. A clutch consists of one disc member driven by the engine, a member which is connected to the transmission and a friction plate which links the two, together with springs and withdrawal mechanism.

On the Reliant, the flywheel rotates with the engine mainshaft. The gearbox shaft carries the friction plate, which is sandwiched between the flywheel and its spring-loaded pressure plate. Thus when the clutch is driven by the crankshaft it turns as a unit, owing to the pressure exerted by the springs, the friction between the plain surfaces and the lined disc being such that the clutch can transmit the drive to the gearbox.

When the withdrawal mechanism is operated the pressure of the springs is relieved. The flywheel and the pressure plate fixed to it still revolve, but the friction between these and the disc is now too low to transmit movement. The disc therefore remains stationary, so no drive is transmitted.

By progressively releasing the withdrawal mechanism the revolving plates can be brought into gradual contact with the disc. At first they slip, but as contact is increased the disk speeds up until with the full spring pressure restored the whole clutch is once again rotating as a complete unit. This is what happens each time the car moves off from a standstill, or when the driver changes gear.

THE STEERING

When a Reliant is driven along a road it goes in a straight line because —although the fact is not immediately apparent—the front wheel is trailing, rather like the castor of an armchair. The characteristics of the steering depend to some extent upon the amount of *trail* specified by the designer, and to some extent upon other factors. One of these is the *castor angle*—the angle at which the king pin carrying the wheel is set— and others are the weight distribution and the position of its centre of gravity.

The manner in which the suspension systems act plays a great part in determining whether the car handles well or not. The Reliant design utilizes a pivoted front arm and a live rear axle suspended on semi-elliptic springs. In each case movement has to be *damped*—if there were no damper the spring would thrust the wheel up and down with a rapid action and so cause the vehicle to pitch.

To prevent this, a hydraulic damper is used. This consists of an oil chamber and a disc valve. This valve is so designed that when the sliding member rises it permits the oil to pass through with little or no resistance. On the return stroke the valve is partially closed and this slows down the rate at which the oil can return to the chamber in the sliding member.

THE BRAKES

Just as important as making the car move is the ability to make it stop. This is the job of the brakes, which are of the internal-expanding type. Each wheel carries a drum, the inside surfaces of which are accurately ground so that the drum is completely round and true. Closing the drum is a *backplate*, and affixed to this plate is a *pivot pin*. Diametrically opposed to the pin is a cylinder and piston, connected to the brake pedal by hydraulic lines. Two *brake shoes*—semi-circular in shape, with a friction lining rivetted to the outer curve on each shoe—are fitted with one end of each shoe butting on the pivot pin and the other on one face of the pistons. They are held together by springs and the whole backplate assembly is fixed rigidly to the vehicle.

When the brake pedal is operated the action of the master cylinder piston pressurizes the hydraulic system and the pistons in each wheel cylinder move and press the free ends of the shoes outwards. This brings the friction linings into contact with the inside surfaces of the drum, decelerating the car.

A brake is basically a form of heat-exchanger. The friction created by the linings rubbing on the surface of the drum absorbs energy which would otherwise by devoted to driving the vehicle, and this energy is converted into heat which is dissipated from the surface of the drum. An auxiliary hand control works through cables and rods on the rear brakes only.

3 The Tool Kit

It is virtually impossible to make a bigger mistake when setting out to maintain or overhaul a Reliant than attempting to do the job with inadequate tools. To carry out even routine maintenance jobs properly calls for the use of a good-quality tool kit, while major overhauls can quite often require the use of special tools designed by the manufacturer to do one specific job and one only.

Each car is equipped with a tool kit upon delivery, but with the exception of one or two items this is intended to act only as a stand-by kit capable of dealing with road-side emergencies but very definitely not suited to sterner work. In addition, the use of special service tools for certain jobs is not dictated simply by lack of intelligence on the part of the designer, nor because the makers can sell such tools at a profit. It is dictated by the very fact that the mechanical parts are made to precision-engineering standards and that some of the assembly work is such that once the components have been assembled at the factory only special tools have the slightest chance of freeing the various parts concerned.

Even where the jobs to be tackled do not call for the use of special tools they will still require the use of *good* tools. Cheap tools are a bad investment, for not only do they not wear as well as they should but they also have an infuriating habit of ruining nuts and bolts.

The first essential is to buy a really good set of chrome-vanadium *open-ended spanners* in A/F sizes. A set of half a dozen double-ended spanners will give a range of sizes sufficient for most of the work, and will cost only a couple of pounds. For side-valve models, buy Whitworth spanners, for A/F nuts occur only at a few points—on the engine/gearbox flange, for example.

Next it is essential to have a set of strong *box spanners* or, even better, a set of *socket spanners*. And I would not be without my set of *ring spanners*, which—though less handy in confined spaces than are open-enders or sockets—enable a very good grip to be obtained with no danger of the spanner slipping.

Almost equally essential are a pair of really good *screwdrivers*, with insulated handles. One screwdriver with a five-sixteenths of an inch blade and an electrical screwdriver with an eighth of an inch blade are the minimum requirements.

You will also need a pair of *pliers* equipped with wire-cutters. Pliers are indispensable for electrical and cable work. A nest of small B.A. spanners is also useful for electrical jobs. And you will require feeler gauges for gap-setting the tappets, plugs and contact points.

USING YOUR TOOLS

There is really rather more to using even the simplest hand tools than merely placing them in position and tugging hard. Each particular type of spanner has its own characteristics, and each is better suited for one type of job.

The great all-rounders are unquestionably the open-enders. These can be used in confined spaces, and they have the great advantage that the jaws are angled. This means that an open-ender can be used to loosen an awkwardly-situated nut and then, when the limit of its movement has been reached, reversed to give fresh purchase. In this way it is possible to undo a recalcitrant nut by easy stages.

It is, of course, essential that only the right size of spanner should be used. The open-ended spanner is designed to apply its pressure on the flats of a nut or bolt, and it is consequently made with jaws of just the right width to grip its bolt. If a larger spanner is used, instead of gripping the flats the jaws will press against the angles of the bolt. Then one of two things can happen. Either the spanner gouges away the angle of the bolt, giving a rounded head which no ordinary spanner could grip thereafter, or else the bolt head slightly springs the jaws of the spanner, which is promptly ruined.

Damage to the jaws can also be caused by applying excessive force when trying to free a bolt which refuses to budge. There is a temptation, under these circumstances, to slip a piece of piping over the free end of the spanner to increase the leverage. Although this is permissible where due care is used it is all too easy to apply excessive force and to spring the jaws of the spanner or snap the bolt or stud.

Ring, box, or socket spanners are at a great advantage when really obstinate nuts have to be dealt with. Ring and socket spanners both grip on the angles, not on the flats. They are consequently able to apply pressure at half a dozen points where the open-ended spanner can do so only at two. Box spanners, providing they are stoutly constructed, go one stage better. A box spanner applies force at both angles and flats, all the way round the bolt. Frequently, however, the tommy bar used to turn the box spanner simply bends under the strain, or else the amount of offset between the part of the spanner holding the bolt and the tommy bar hole where the pressure is applied tilts the spanner which then rides off the hexagon.

When using a spanner to tighten nuts or bolts it is important to remember that too much force should not be used. Spanners are made long enough to ensure that, for any given size of bolt, mere hand pressure applied through the full leverage of the spanner will tighten the bolt adequately. If excessive force is used the actual material of the bolt can be weakened sufficiently to cause it to fracture. This point should also be borne in mind when tightening bolts which are threaded into light alloy. Here the steel bolt is much harder than the material into which it is screwed, and over-enthusiasm with the spanner can easily strip the thread inside the hole.

Pliers, of course, should never be used as a makeshift spanner since the jaws can never be parallel and the serrated grip is almost perilously liable to slip. A rounded hexagon is the inevitable result if it does.

Adjustable spanners should never be allowed near the car. These are a butcher's tool, not a mechanic's. Here again it is impossible to align the jaws with sufficient accuracy to enable a satisfactory grip to be obtained.

Screwdrivers should have their blades properly ground so that, in side view, the blade is first concave and then runs parallel to the tip. This enables the blade to seat itself properly in the slot and to apply pressure which is evenly distributed. A screwdriver whose blade is wedge-shaped in side view does not seat properly, and instead of an even pressure on the sides of the slot it exerts all its force on the edges which, understandably tend to crumble under the strain.

After use all tools should be wiped clean. They should be kept in a dry place, protected from dust by being wrapped in rag; and if they are used fairly infrequently they should also be very lightly oiled. This light film of lubricant, of course, must be wiped off before they are again put to use.

4 Fault Tracing

WHEN a doctor wishes to diagnose a patient's illness he works methodically, listing the various symptoms to build up an overall picture of the complaint. This done he can identify it and give treatment accordingly. Exactly the same type of diagnosis has to be made in the case of an engine which refuses to work. Obviously there is a fault—some reason why it will not work—and before the fault can be cured it has got to be located and identified. The search for it must be just as methodical as the doctor's approach.

If certain requirements are being properly fulfilled the engine *must* work. If it is not working it follows that one or more of these requirements is not being met. Fault tracing boils down to discovering which of them it is.

An engine must work if the correct charge of petrol/air mixture is being induced into the cylinders at the right time, properly compressed, fired at the right moment, and the residues properly exhausted. The first stage in checking must, therefore, be the obvious one of ensuring that there is, in fact, a supply of petrol and that this is reaching the carburettor.

As an invariable first step, always check the fuel tank to make sure that it contains sufficient fuel. Then comes the task of ascertaining that the fuel is reaching the carburettor. Flowing of the fuel could be prevented by a blockage in the lines, a faulty pump, a choked filter, or a jammed needle valve. Detach the fuel pipe at the carburettor and operate the hand lever on the fuel pump. Petrol should spurt out. If it does the chamber can be opened up for inspection. First, however, reconnect the pipe.

When the chamber is open, the pump may be operated again so that you can see whether or not the fuel is flowing through the needle valve properly. Also look for signs of dirt in the chamber itself. If there is sediment on the bottom take the opportunity of swilling the chamber out thoroughly before everything is replaced.

Normally, this initial check on the fuel system will have taken only a matter of a few minutes. It will have given one of two answers: either that fuel is reaching the float chamber, or that it is not. If it is not you have found at least a contributory factor to the breakdown and this should be rectified before proceeding.

FAULT TRACING

Suppose that no fuel flowed through the pipe. That would show that the trouble lies somewhere between the free end of the pipe and the fuel tank—in the pipe itself, in the pump, or in the tank pick-up. In this case the next step will obviously be to detach the pipe at its lower end and then to operate the pump. If fuel then flows the pipe will be isolated as the seat of the trouble. If not, then the obvious inference is that the pump filter itself is blocked with dirt or that the diaphragm is faulty. So the filter dome will have to be removed and the filter and pump checked.

It is possible for the fuel system to be at fault by supplying too much fuel as well as by supplying too little. Overflooding, as this form of trouble is called, is easily recognizable. Fuel drips from the carburettor, and the engine—if it runs at all—constantly misfires and sounds distinctly "lumpy."

Unless pump pressure is wrong only the float assembly can be responsible for this particular form of trouble. The float may have been punctured, in which case it would simply sink to the bottom of the chamber and allow the valve to remain open. More likely, however, is the ingress of dirt into the valve assembly. Even a tiny speck of hard matter is sufficient to prevent the needle seating properly, and this keeps the valve partially open. The effects of this milder form of overflooding would be more noticeable at low engine speeds where the excess fuel was not being used up quickly enough than at high engine speeds. Finally, the needle itself may have become bent. In each case, thorough inspection of the float assembly is the only way of pinpointing the exact cause.

Where the initial inspection of the fuel system shows no immediately obvious fault the next stage of the fault tracing should be switched to the ignition system. First of all remove the sparking plugs and inspect the gaps. Where inspection and checking with a feeler shows that the gap is neither too closely nor too widely set each plug in turn should be reconnected to its H.T. lead and then the metal body of the plug must be placed in contact with the cylinder head or some convenient metal part of the car in such a position that the spark gap can easily be seen. With the ignition switched on the engine is turned over. As this is done a good fat spark should jump across the plug points. This check should be repeated several times, and if no spark results a brand-new plug—an essential "spare" which should always be carried—should be substituted and the check repeated. If the new plug sparks and the old one didn't then the obvious inference is that the old plug's insulation has broken down, and fitting the new plug in its place should cure the trouble. If, on the other hand, the new plug fails as well the trouble lies somewhere between the sparking plug terminal and the battery, and a much more exhaustive check is needed.

Complete engine failure for any reason other than ignition or fuel trouble is unlikely, save in the somewhat remote event of such a vital part as the timing gear drive to the camshaft being stripped. Other troubles are, therefore, likely to show themselves in reduced performance or erratic running.

One of the likelier causes of lack of pulling power is the tappet setting being incorrect. The tappets are given a slight clearance to enable expansion to take place without adverse effects on the valve seating. If, however, a tappet is set too tight, expansion as the engine warms up will tend to hold the valve away from its seat slightly and thus reduce the compression in that particular cylinder. It is possible, where this is suspected, to deduce where the fault lies from the way in which the engine behaves. If an inlet valve is not being properly closed there will be a tendency for the engine to "spit back" through the carburettor, since some mixture will be driven back during the compression stroke. Where an exhaust valve is not seating properly the mixture tends to be driven into the exhaust system and to be ignited there by the heat, giving a constant banging or rumbling in the exhaust pipe.

A rough check can be made thus: with the engine running detach each ignition lead in turn; when the lead is taken off the cylinder which is *not* developing its full power the engine speed will fall only slightly, but when the lead is removed from a cylinder which *is* contributing its full power the fall in engine speed will be marked, and the unit may even stop.

It is not possible, of course, to reset the tappets accurately when the engine is hot, but since damage to the valve seats can result from running with a tight tappet it is possible in an emergency to slacken off the wrongly-set tappet until there is just a barely-perceptible amount of play. This, of course, is only a get-you-home measure, and once the engine has cooled the tappets must be reset in the normal way.

Loss of compression can be caused in several other ways, although none of them is at all usual. Distortion of the cylinder head joint, which could result if the engine is run when suffering from chronic overheating, is one of them. This, obviously, calls for workshop treatment. In such a case there might be a distinct hiss of escaping gas audible at the joint all the time the engine was running, and since air would be induced into the cylinder, thus diluting the mixture, the engine would also tend to overheat. A leakage of coolant from the broken joint might also be noticeable.

Following a seizure of the engine—either through working it too hard before it was run in or from failure of the lubrication system, piston rings might be fractured. Besides losing compression the engine would also begin to take oil into the combustion chamber. This oil, of course, would burn and the resulting smoke would issue from the exhaust pipe. And when an engine loses power and smokes after a seizure the only wise course is to stop immediately, since the rings are almost certainly broken. Any further running might easily cause deep score marks on the bores —where the broken ends of the rings act as efficient cutting tools—and the engine might then be virtually ruined.

Exactly the same process of elimination that is used for locating mechanical troubles is employed when tracing faults in the lighting system. Faced with electricity, of course, most laymen give it best first time, but in actual fact electrical work is reasonably straightforward providing that the magic word "circuit" is borne in mind. Circuits are, in fact, the key to electricity. If electricity is present and the circuit is complete the current *must* flow through it. If electricity is present but is not flowing it follows that the circuit is not complete.

Faulty circuits are of two types—the open circuit and the short circuit. In the first case there is a break in the circuit and the wires on the far side of the break, viewed from the electrical source, are dead. In the second case the current is still flowing but is following a shorter path to earth—as would be the case, for example, if one end of a live lead had become detached from its terminal and had earthed itself on the chassis.

Obviously the first essential is to be able to understand a wiring diagram. At first sight this may appear rather disconcertingly like a plan of a rather complicated suburban railway—and oddly enough it is not at all a bad idea to regard it in this light. If the various leads are thought of as railway tracks the whole idea becomes enormously simplified. It is as well to remember, however, that one important main line is not shown. This is the earth return. One terminal of the battery is connected straight to earth, and all the components are similarly earthed. This, therefore, forms one complete half of the circuit.

To trace, for example, the circuit which lights the main headlamp bulb filament one takes as the starting point the unearthed terminal of the battery and follows the lead shown on the wiring diagram leading from this terminal. It goes to a point on the lighting switch. From the switch another lead is taken, through the dip switch, to the headlamp bulb, and thence to earth. If, therefore, the tumbler of the lighting switch connects these two main switch terminals there will be a complete circuit, for after passing through the filament the current returns via earth to the battery. Sometimes, where a rather more complicated circuit is involved, it helps to trace it out individually on a sheet of paper.

Having found the circuit, the next job is to check it. Obviously, the

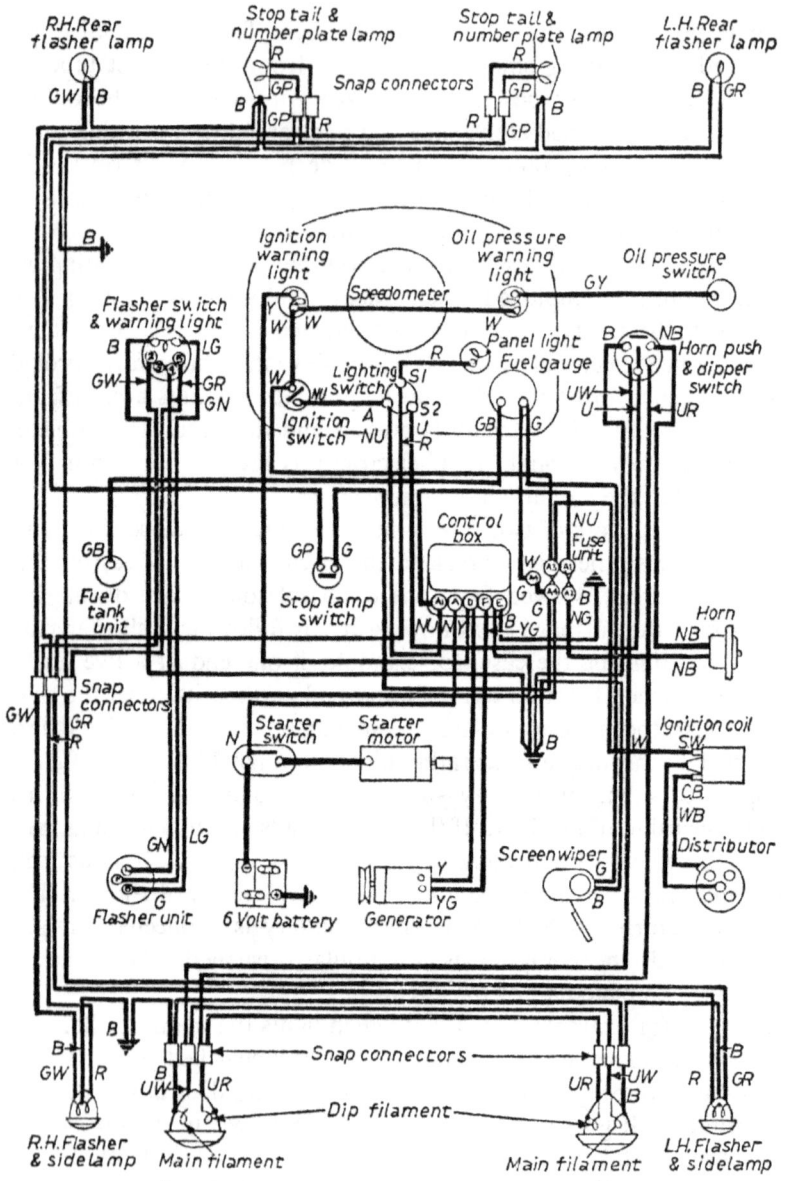

Fig. 4. Wiring of the Regal Mark 3 (6-Volt System)

On this diagram, and on Figures 5 and 6, the following code is used: *B*, black; *U*, blue; *N*, brown; *G*, green; *P*, purple; *R*, red; *W*, white; *Y*, yellow; *L*, light; *LH*, left-hand; *RH*, right-hand.

first stage is to find out whether or not electricity is present, and here a spare bulb in a holder with a "wander" lead attached can be pressed into service. First, connect the bulb across the battery terminals. If the battery is in order the lamp will light. Once certain on this point it becomes necessary to check each individual lead in the circuit in question. This is made considerably easier by the fact that each wire is given an outer casing of a different colour.

In the circuit used here as an example the next step would be to lift out the lighting switch, complete with its leads, and to apply the lamp lead to the appropriate terminal while earthing the holder. Once again, the lamp should light, indicating that electricity is flowing as far as the terminal. The switch should then be set to the appropriate position and the check repeated on the second terminal. If that passes muster the next stage of the wiring—from the light switch to the dip switch—should be similarly checked. Finally the one remaining lead—from the dip switch to the headlamp bulb—should be tested. This assumes, naturally enough, that one has examined the headlamp bulb first, to make sure the filament is intact, and has checked that the earth loop wire to the chassis of fibre glass bodied cars is properly attached.

When the faulty section of wiring is located it should be closely examined so that the exact cause of failure can be ascertained. A short circuit can often be detected by shaking the wire and listening carefully. As it contacts metal, the characteristic crackling of current short-circuiting can easily be heard. A break in the wire, concealed by the outer insulation, can usually be found by holding each end and pulling gently. If the inner cable has broken the lead will stretch at the point at which the fracture has occurred.

Where the suspect lead is a very long one and is inaccessible an alternative method of checking is to by-pass it with a temporary external lead. This is connected to the terminal at each end, and the switch is then operated. If the hitherto inoperative component—a tail light, for instance—works when thus connected it shows that the fault is in the lead. In some cases it is possible to draw a new lead through a conduit by using the old lead as a guide. The new lead is securely fastened to the old one—by wiring the terminals together, for example—and the old lead is then withdrawn, pulling the new lead into position.

When repairing fractured leads it is important to ensure that no undue electrical stresses are set up and that the insulation is made good. All joints should be twisted together neatly, and well wound with insulating tape to make leakage impossible. Where terminals have been undone they must be done up again tightly, since a loose terminal can cause

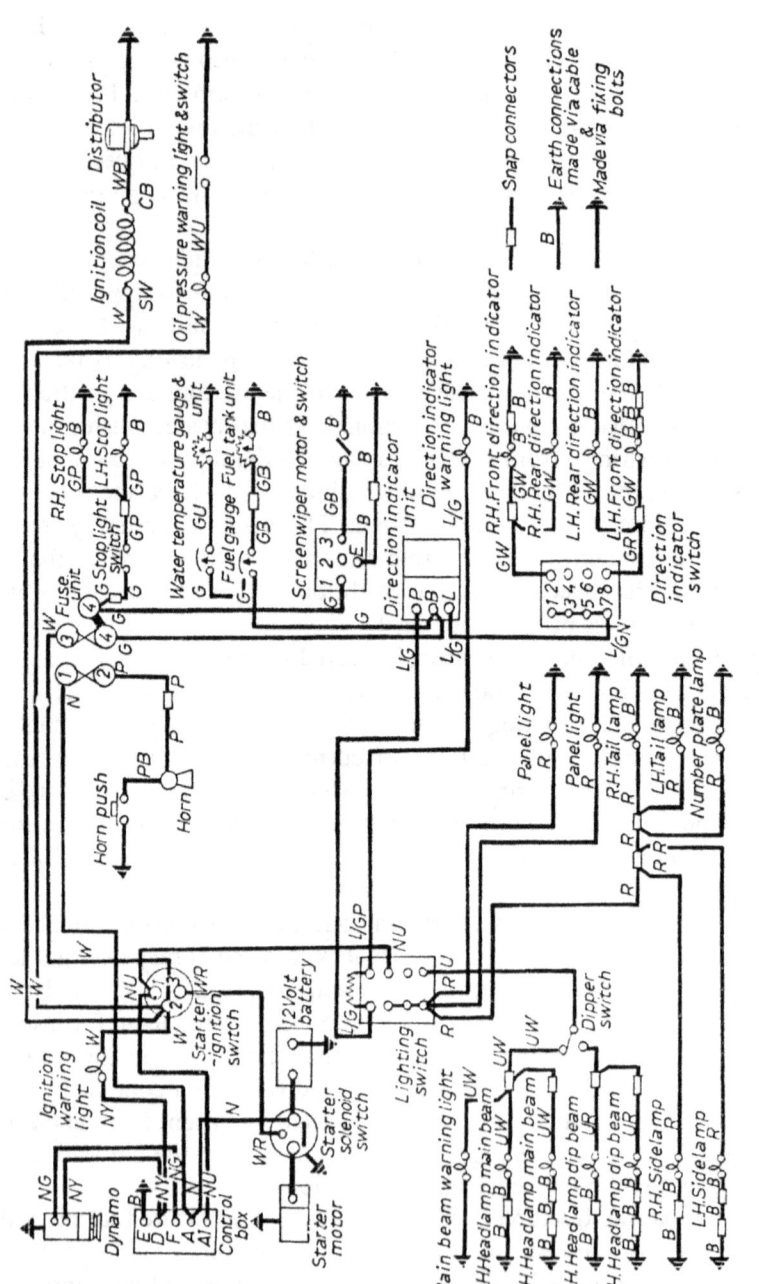

(a)

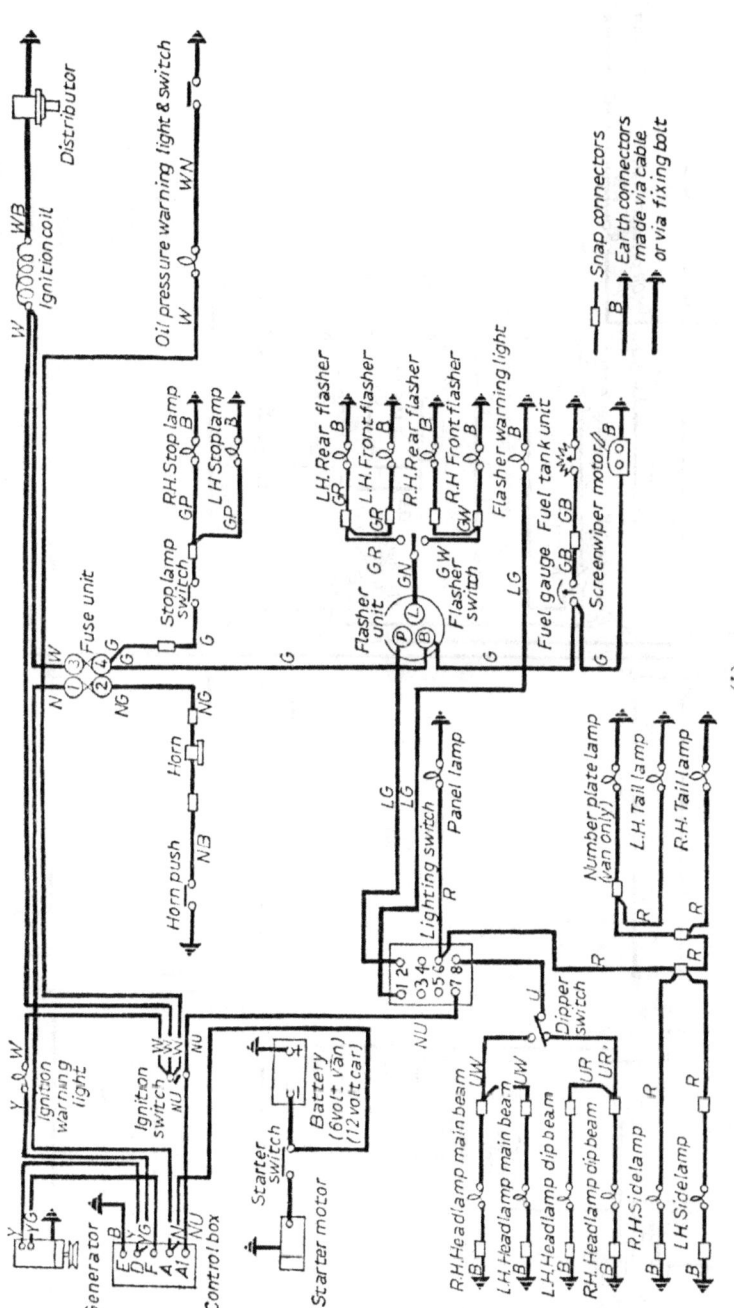

Fig. 5a and 5b. Wiring of the Regal Mark 3 and 4 (12-Volt System)
For coding see the caption to Figure 4.

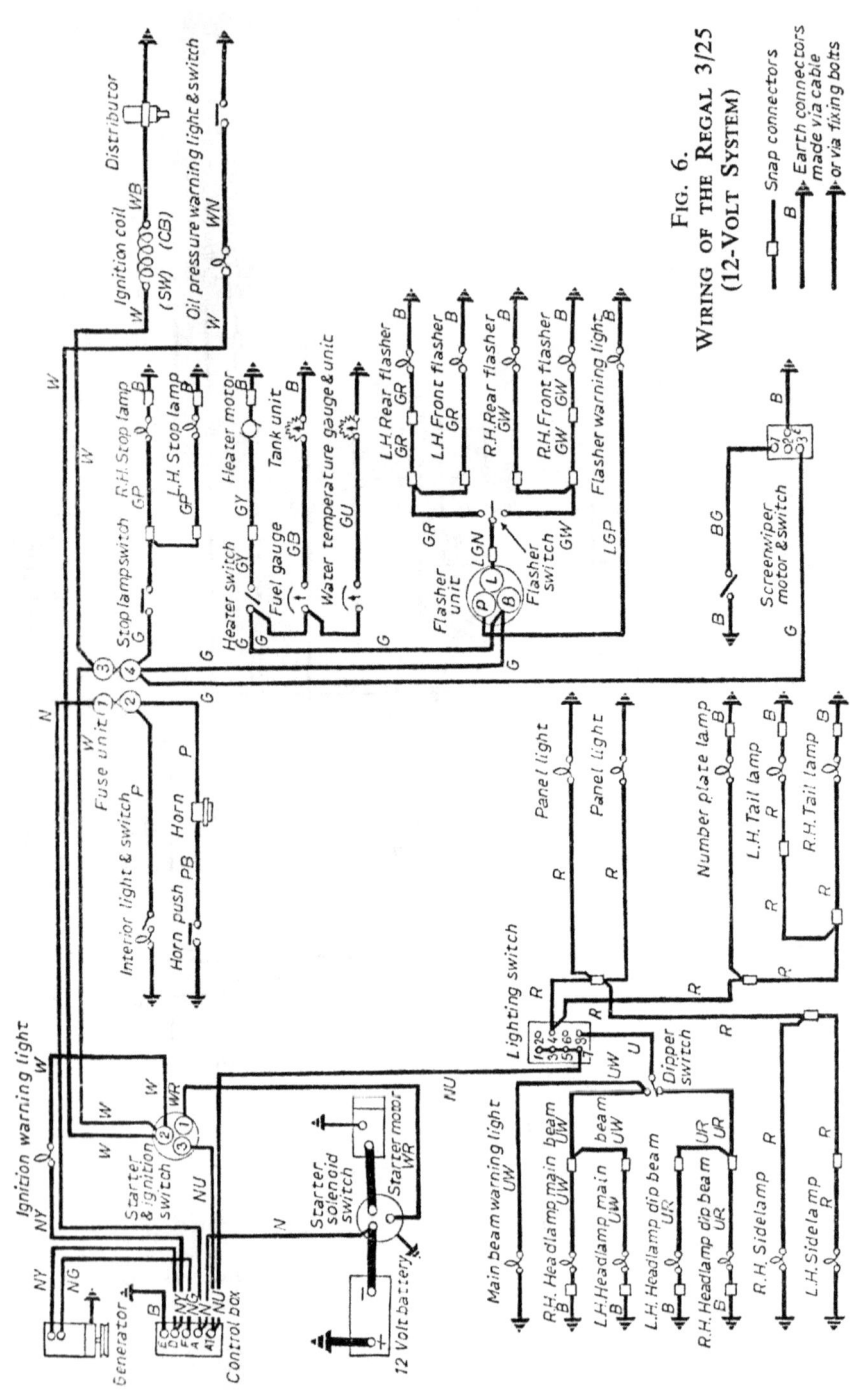

Fig. 6.
Wiring of the Regal 3/25 (12-Volt System)

electrical stresses through spasmodic breaking of the circuit. Where a soldered joint has failed it is essential that it be remade with solder, and not merely taped.

With these pointers borne in mind, there is no reason why the ordinary owner should not be able to trace most faults occurring in the engine and in the electrical systems of his Reliant, and provide at least a temporary cure for them.

5 Routine Maintenance

OBVIOUSLY there is quite a difference between routine maintenance—the day-to-day adjustments and minor repairs which all vehicles need—and major overhauls. But both have their part to play in keeping your Reliant in good order and ensuring for it as long a working life as possible. In fact, I'd rate day-to-day maintenance as the more important of the two. A Reliant repays constant and sympathetic attention to its everyday needs—but no car is ever one whit the better for being constantly stripped and tinkered with. Preventive medicine, not surgery, is usually the best treatment for any complex piece of mechanism. Yet many owners fall into the error of failing to give their vehicle regular minor attention while ceremoniously pulling them apart year after year. That is the exact opposite to the correct approach.

Properly driven and well maintained a Reliant should cover a good 40,000 miles before the engine need be stripped. Up to that time only two decokes should have been needed. But if the routine maintenance has been neglected or skimped the overhaul mileage may be drastically shortened and its cost increased.

The reason is simple enough. Maladjustments have a cumulative effect. Little enough harm is likely to result, for instance, if a car covers a few hundred miles with a tight tappet. But if thousands of miles are covered in such a condition because the owner forgot to check the clearances at regular intervals all sorts of trouble may result. The hot exhaust gases will gradually burn away the seating of the valve until, at last, compression and ignition pressures are so reduced that that cylinder is never operating at anything like its full power. Fuel consumption rises and the car's performance falls off. And when at last the fault is discovered you find that adjustment will no longer provide a cure. Instead you need to have a top overhaul—with perhaps a new valve and recut valve seat—to make good the result of neglecting to make a regular tappet check. Is it worth it?

Take the brakes. Here deterioration is constant, but so gradual that you tend to adjust yourself imperceptibly to the lessening power of the brakes in everyday driving. But emergencies *don't* adjust themselves—and comes the day when you need to stop quicker than you've ever

ROUTINE MAINTENANCE

stopped before. And what do you find? Why, that brakes which six months before would have done it easily are now no longer efficient enough. This lesson can be effective—providing you survive to appreciate it. Yet a set brake test carried out once a week as a matter of routine would ensure that you *know* what the condition of the "anchors" is.

TASK SYSTEMS

Constant and methodical inspection is the best way of preventing any such troubles but the usual recommendations—based on elapsed mileages—are difficult to carry out if a full log of work done is not kept. This was a problem which the Services had to face long ago, and one answer was the Task System. This called for daily and weekly checks as a matter of routine.

Modified, such a system is ideally suited for a privately-maintained car. And here I've detailed two routines—one daily and one weekly—either of which could be employed by the average Reliant owner. They cover all the more essential check points, and to carry them out need take no more than, on average, ten minutes a day or an hour each weekend.

Daily Task System

Sunday. Check adjustment of brakes; check hydraulic fluid level; check action of hand brake; check security of nuts and bolts and unions in braking system; check lubrication of brake linkages; test security of back plates.

Monday. Check oil levels in sump and transmission; check coolant level in radiator; check all controls for adjustment and free movement; check tyre pressures.

Tuesday. Check all exposed electrical wiring for signs of abrasion or fracture; check all terminals for security; check operation of lamps.

Wednesday. Examine tyre treads and remove any trapped stones; check walls for cracks; test tyre pressures; check wheels for security; check steering backlash.

Thursday. Check action of clutch; check brakes for distance to stop from 30 m.p.h., using known reference points.

Friday. Check all nuts and bolts for security; rock car to test dampers.

Saturday. Listen to tick-over to detect tappet noises, and adjust if necessary; check contact-breaker gap; check plugs for condition and gap; check fan belt.

ALTERNATIVE WEEKLY SYSTEM

Week 1. Check adjustment of brakes; check hydraulic fluid level; check action of handbrake; test security of nuts and bolts and unions in the braking system; check lubrication of brake linkages; test security of back plates; test brakes for distance to stop from 30 m.p.h., using known reference points; check tyre pressures.

Week 2. Check oil levels in sump and transmission; check coolant level in radiator; check all controls for adjustment and free movement; examine tyre treads and remove trapped stones; examine side walls; check tyre pressures.

Week 3. Check all exposed electrical wiring for signs of abrasion or fracture; check all terminals for security; check operation of lamps; check tyre pressures; check oil and coolant levels.

Week 4. Check all nuts and bolts for security; rock car to test action of dampers; check steering backlash; listen to tick-over to detect tappet noises, and adjust if necessary; check contact-breaker gap; check plugs for condition and gap; test fan-belt play; check tyre pressures.

By employing such an approach to routine maintenance you will ensure that most of the major points are checked, by the daily system, at least once a week. Even allowing for a pretty substantial utilization of the car this should mean that no fault could go undetected for more than 500 miles, and most defects would be discovered almost before they have had time to develop.

With the weekly system, of course, there is a greater time-lag—though one or two points are double-checked during the period. Where the car is used mainly as a weekend runabout the time-lag does not assume a great significance, but if the monthly mileage is much in excess of 500 then it would be better to make the checks on a daily rather than a weekly basis.

Note, though, that the idea is to *check* the relevant items, not necessarily to carry out any adjustments. Where the routine examination discloses no fault then no actual work need be carried out.

LUBRICATION

Obviously, the system of checks does not take into account either seasonal or periodic jobs such as oil changes and greasing. These must still be done on an elapsed mileage basis, and a record should be kept of the date and the mileage reading at which each service was made.

As memory can be notoriously fickle it is no bad plan to have such a reminder actually on the car itself. I find that one of the best methods is to stick to the underside of the bonnet a small strip of self-adhesive

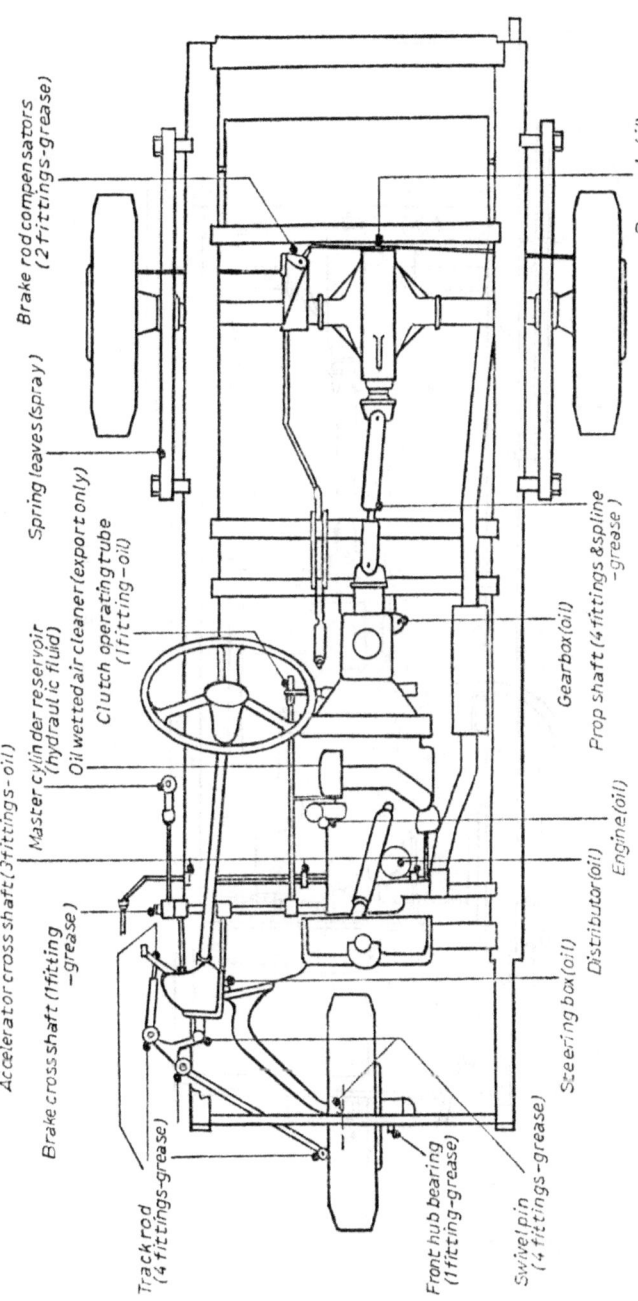

Fig. 7. Chassis Lubrication, Side-Valve Models

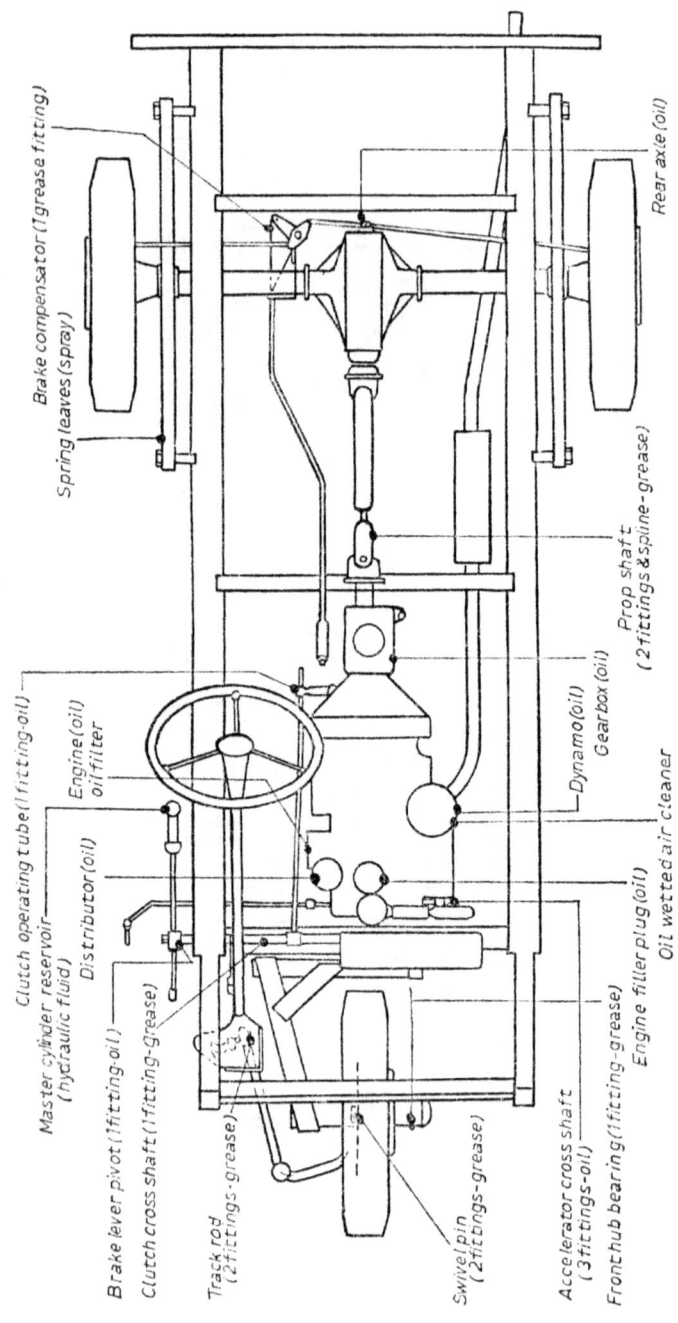

Fig. 8. Chassis Lubrication, Regal 3/25

coloured tape, which is bound to catch the eye. On this, jot down either the date and mileage at which, for instance, the oil was last changed; or the mileage at which it is next due for a change. But be sure to use just the one system and stick to it.

It is a great mistake to neglect these periodic lubrication sessions. The engine oil, contained in the sump, is constantly circulated through the power unit. In time, it becomes dirty—contaminated with minute pieces of metal worn from the pistons, rings, and bores, and mixed with water and petrol which percolate down from the cylinders. Remember that for each gallon of fuel burned in the engine rather more than a gallon of water is produced. Much of this finds its way out of the exhaust pipe in the form of vapour—you've probably noticed how much steam is emitted from the tail pipe on a cold morning just after the engine has started. But some of it condenses on the cylinder bores and runs down the walls into the sump. Left to itself, this will form sludge. And, of course, the net result of all this is to lessen the oil's lubricating properties.

How to Change the Oil. It is useless to try to make an oil change with the engine cold, for the dirty oil will not flow properly. Instead, let the unit warm up first—better still, go for a few miles' motoring first—and then place your drain tray under the unit. Take off the filler cap to obviate any suction effect. Remove the sump drain plug and the oil will gush out, carrying with it most of the impurities contained in the sump.

Replace the drain plug and pour in flushing oil. Run the engine for a further five minutes—at a standstill this time—to circulate the oil. This clean light lubricant will wash away any deposits in the oil channels and galleries.

Stop the engine, allow the flushing oil a few minutes to drain back into the sump, and then remove the drain plug again and leave the car for a while so that all the flushing oil runs away.

Where an oil filter is fitted (on 3/25 models and on side-valve engines later than No. 18278) renew the filter. Then, with the car level pour in fresh engine oil until the correct level on the dipstick is reached.

Check the oil level again after the first trip which you do, and top up if necessary. Do not mix different grades of oil, and don't make experiments with your lubrication. The grades of oil specified in the Appendix to this book are those arrived at after much careful experimental work on the part of the factory and the oil companies. You will not, believe me, be able to improve on their results—I have seen for myself just how painstaking the oil laboratories are on this point—and you may easily take thousands of miles off the life of your engine by trying to do so.

Oil Changes, Gearboxes. Though the oil in the gearbox is not subjected to the same extremes as is the engine oil it cannot be expected to last for ever. And, remember, very heavy strains are imposed on the thin oil film between the meshing teeth of the gearwheels.

Gearbox oil changes are made in just the same way as are those for an engine. Take the car for a run to heat the oil. Place the drain tray into position and remove the drain plug and the filler plug. Let the old oil drain thoroughly, then refit the drain plug and top up with fresh oil to

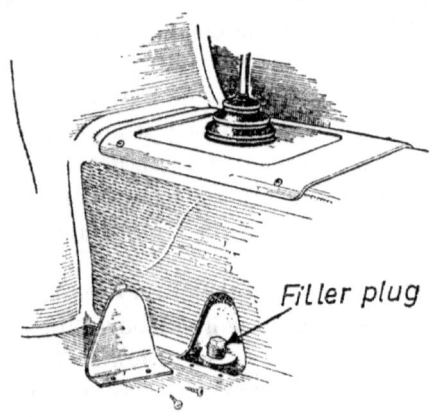

FIG. 9. GEARBOX FILLER

A small plate held by a pair of screws conceals the filler plug. Later models have a dipstick and filler under the large lid surrounding the gear lever, reached through a flap in the carpet.

the level of the filler plug. Do this at intervals of 15,000 miles—or annually if your year's mileage is under 15,000.

Oil Changes, Rear Axle. At 15,000-mile intervals drain and refill the rear axle. The procedure is the same as for the gearbox, but as heating-up in the axle is slow it is best to do this job after a long day's run if possible.

General Lubrication. Besides these regular oil changes, there are a number of greasing and lubrication points on the car which need periodic attention. For these, consult the charts given on pages 31 and 32.

Sump Oil Gauze. Both side-valve and overhead-valve power units have a strainer gauze in the oil tray which requires cleansing at intervals of 15,000 miles, or once every eighteen months on an average mileage. To do this drain the oil as described above, and then remove the sump. With clean petrol brush the oil gauze clear of dirt and wash the sump

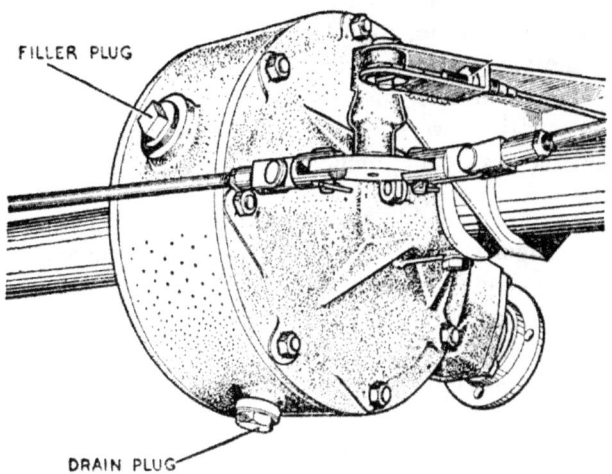

FIG. 10. REAR AXLE HOUSING, SIDE-VALVE CARS
Besides the axle the brake linkage should be periodically oiled.

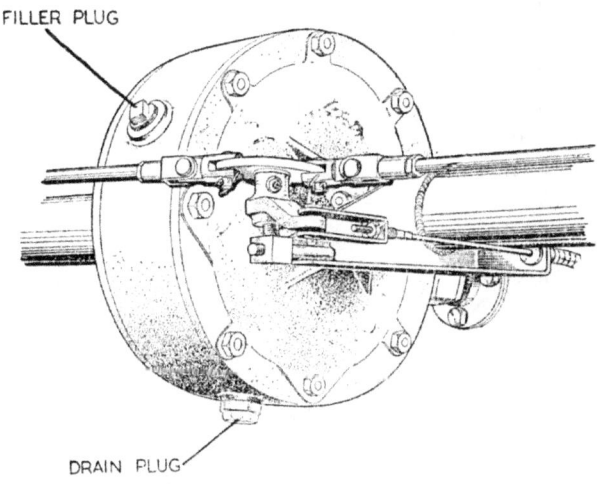

FIG. 11. REAR AXLE HOUSING, REGAL 3/25
Though the axle plug positions are as on the earlier cars, the brake linkage differs.

thoroughly in a petrol bath. Then invert it on clean newspaper and allow it to drain and dry thoroughly before refitting.

Remake the sump joint very carefully using a new packing washer. Tighten all sump bolts to finger tightness first and then take up each a few threads at a time, working diagonally from bolt to bolt until all are tight. Insert the drain plug, tighten it, and refill the sump with fresh oil.

TAPPET ADJUSTMENT

Side-Valve Engines. Since accessibility is limited, there is a temptation to overlook tappet adjustment on this unit. It should, however, be checked whenever a noise test suggests that one or more of the gaps has opened up, or where inspection of the sparking plugs indicates unsatisfactory operating conditions in one or more cylinders. Adjustment should be made with the engine cold.

The tappets are located behind a plate on the left-hand side of the cylinder block. Remove this and take out the sparking plugs. Then place an open A/F spanner on the large nut securing the engine pulley to the crankshaft and rotate the shaft until the No. 1 valve (exhaust) is fully open—that is, raised as high as it will go. You can now check the clearance on No. 8 (exhaust) valve using a feeler gauge. The correct clearance is 0·004 inch. Insert the appropriate feeler into the gap. It should slide in easily. Then try the next size gauge—probably a 0·006-inch feeler. This should refuse to fit if your gap is correct. If the correct feeler will not fit, make the adjustment and then check with the larger gauge as well.

The actual method of adjustment is to loosen the lower lock-nut on the tappet and to screw the tappet itself either nearer to or further from the valve stem until the right clearance is obtained. Then hold the tappet and tighten the lock-nut, finally re-checking the gap.

Check in the following sequence—

Open No. 1, check No. 8   Open No. 5, check No. 4
Open No. 2, check No. 7   Open No. 6, check No. 3
Open No. 3, check No. 6   Open No. 7, check No. 2
Open No. 4, check No. 5   Open No. 8, check No. 1.

Of these valves, Nos. 1, 4, 5, and 8 are exhausts and Nos. 2, 3, 6, and 7 are inlets.

When you are satisfied with the settings replace the valve chest cover, using a new cord sealing gasket if necessary. Position the breather holes in the case to the top.

Tappet Adjustment, Overhead-Valve Engines. Much the same procedure must be followed with the overhead-valve engine as was detailed in the

ROUTINE MAINTENANCE

previous section for the side-valver. The main differences are that the valves are concealed beneath the rocker cover on the top of the engine; that the tappet adjusters are mounted on the rockers and consist of a

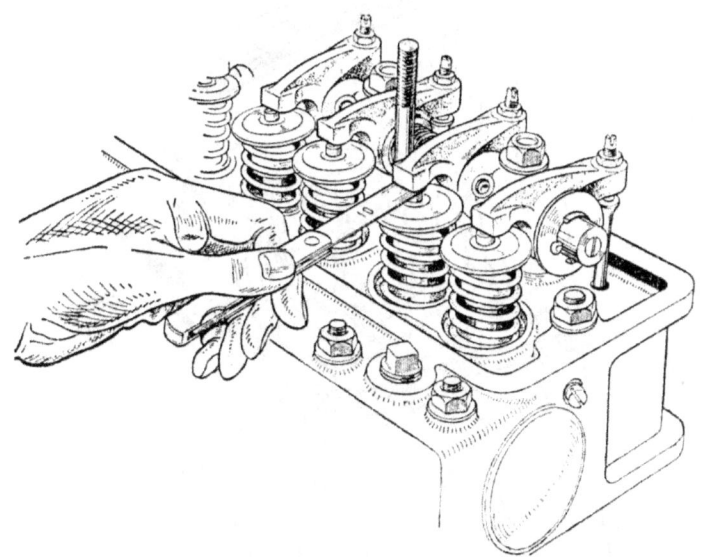

FIG. 12. O.H.V. TAPPET ADJUSTMENT

Great care must be taken when setting 3/25 tappets, and the engine must be absolutely cold if accurate results are to be obtained.

slotted screw secured by a lock nut; and that the correct clearance is 0·006 inch with the engine stone cold (0·010 inch hot).

THE CONTACT-BREAKER

Every 5000 miles (or twice a year at least) the contact-breaker gap must be checked and the points examined. To do this release the two spring clips which hold the distributor cap and detach it. Then grasp the rotor arm and pull it upwards to remove it, so giving access to the points. You will need to turn the engine over, so it is advisable to remove the sparking plugs too. They can be checked at the same time.

Rotate the crankshaft until the contact-breaker points are seen to be fully open. Insert a 0·015-inch feeler gauge. This should just fit. If insertion of the gauge forces the points further apart adjustment will be needed. If, on the other hand, you feel no resistance at all as the gauge is inserted make a second check with, say, a 0·018-inch feeler. Should this be accepted the gap will have to be decreased.

Before making any adjustment ensure that the point faces are neither burned nor pitted. If they are new points may be needed—or at least the

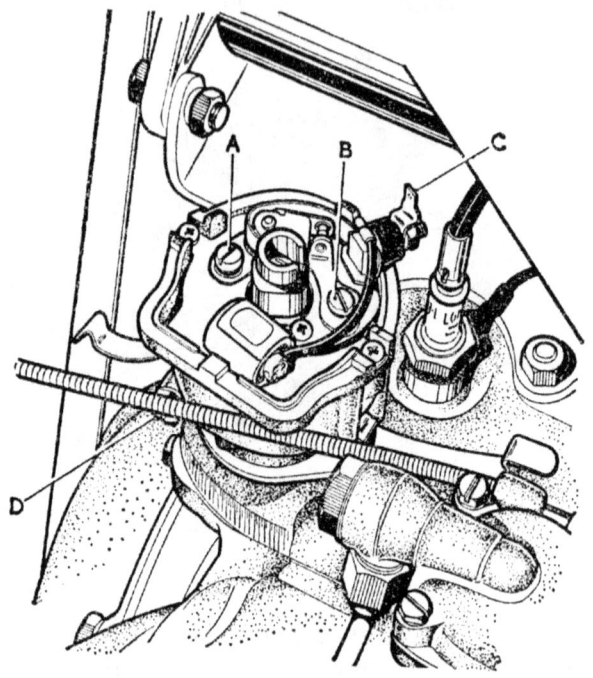

FIG. 13. A TYPICAL DISTRIBUTOR

The screw *A* locks the contact plate. Screw *B* is an eccentric, used to vary the contact-breaker gap. The terminal *C* carries the low-tension current to the distributor which is clamped to the engine by bolt *D*.

existing ones will have to be refaced. How to proceed in such a case is detailed in Chapter 8.

Contact-Breaker, Side-Valve Engines. Where no damage is evident and adjustment is required slacken the front screw on the contact plate sufficiently to allow the plate to move when the adjuster screw at the rear is turned. Rotate this until the correct gap is obtained and then retighten the securing screw. That done recheck the gap to ascertain whether or not it has altered as the securing screw was locked.

Contact-Breaker, Overhead-Valve Engines. The procedure is the same, except that the distributor on these units has only a single screw. Loosen this slightly and move the plate until the correct gap is found. Then retighten the screw and recheck the gap.

DISTRIBUTOR LUBRICATION

Side-Valve Engines. Every 1000 miles insert a few drops of oil into the distributor oiler to lubricate the spindle bearing. Every 3000 miles smear a little grease onto the cam—don't overdo this or some may reach the points and cause burning—and place one single drop of oil on the contact-breaker pivot. Again, be careful not to exceed this amount.

Detach the rotor arm and place a few drops of thin oil on top of the

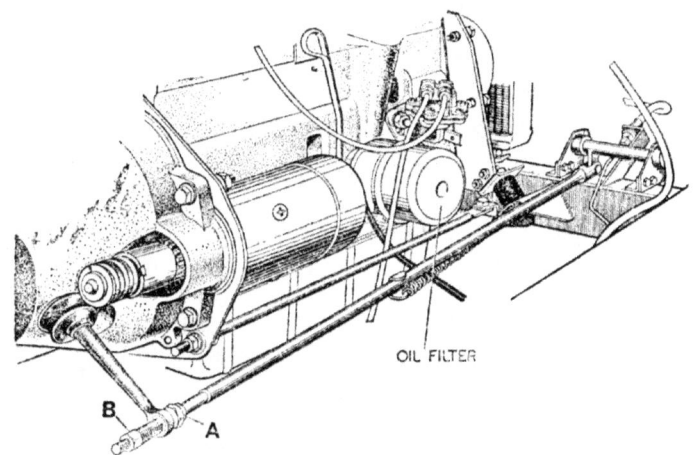

FIG. 14. CLUTCH RELEASE MECHANISM

The clutch is operated by means of an adjustable rod. The layout is very similar on both side-valve and overhead-valve models. This is the 3/25 layout. Note the engine oil filter.

spindle. Don't remove the screw which you will find there; the oil will percolate past its threads to lubricate the cam bearing.

Also at 3000 mile intervals clean engine oil should be used to lubricate the automatic advance/retard mechanism. This is located beneath the contact-breaker mechanism and can be reached by removing the two screws which hold the contact-breaker assembly into the distributor body and lifting it clear. Be certain to refit it in the same position on reassembly.

CLUTCH ADJUSTMENT

Wear on Borg and Beck type clutches takes place only very slowly, and adjustment of the clutch operating control is needed only when play at the clutch pedal becomes excessive. Normally, there should be a minimum of half an inch of free movement before the pedal begins to "bite," and a maximum of around an inch.

To take up excess play, adjust the effective length of the clutch operating rod, which is located under the car on the right side. The layout is depicted in the accompanying diagrams. Loosen the lock-nut and turn the barrel-type adjuster *A* until the right amount of free movement is

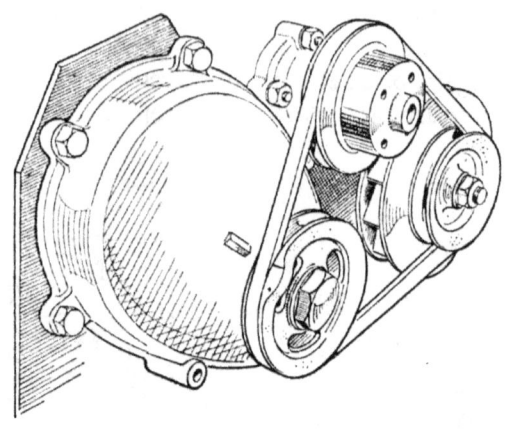

Fig. 15. Belt Drive Layout
On the 3/25 this single belt drives the dynamo, the fan and the water pump.

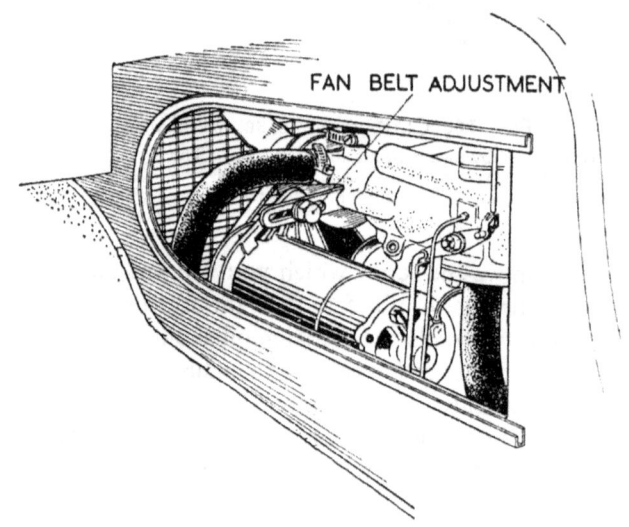

Fig. 16. Fan Belt Adjustment
To tension the belt the dynamo is moved inwards or outwards on its strap after loosening the adjuster.

obtained on the pedal. Then retighten the lock-nut. Finally, turn the nut *B* until the spring tension is just sufficient to hold the adjuster nut firmly in the cupped end of the operating lever.

ROUTINE MAINTENANCE

BRAKE ADJUSTMENT

To compensate for wear in the linings one must alter the position of the brake shoes relative to the drum, and this is done by means of adjusters located on the respective brake backplates. Each rear brake has one adjuster only; but the double-leading-shoe front brake has two adjusters. The method of adjustment is fully detailed in Chapter 11.

Handbrake. This should not require adjustment until the brake shoes are exchanged, and the method of resetting this control is also described under the appropriate heading in Chapter 11.

FAN BELT

Remember to check the fan belt adjustment regularly, for it drives not just the fan but the dynamo and—on O.H.V. cars—the water pump. Set it so that there is half an inch of free play when the belt is tested in the centre. Too much play will cause slip; too tight a belt may snap. The method of adjustment is shown in Fig. 16 on the previous page.

6 Overhauling the Engine

At first sight one can perhaps sympathize with the mechanic who threw up his hands and despairingly exclaimed "To work on this car you need to be ambidextrous and double-jointed!"

True, there is not a great deal of working space around the engine of a Reliant. But there is enough to enable the average private owner to carry out all the routine jobs, and to do a top overhaul without going to the lengths of removing the engine from the chassis. For sterner work—anything which involves operations on the bottom half of the unit—removal is essential.

However, it is the top overhaul—more familiar to most of us as a "de-coke"—which is the work most likely to need doing at home, so let's concentrate on that for a start. Now the object of a top overhaul is twofold. First, it is to remove the carbon deposits which as inevitably build up on the exhaust manifold combustion chamber and piston tops as soot builds up in a chimney. Second, and more important, is to check the condition of the valves and their associated seats and springs.

Now carbon, though generally undesirable, can in fact do a useful job in small quantities by building up a seal around the top of the piston. And its deleterious effects are mainly that excessive deposits can become incandescent, and so cause premature firing of the charge; and also that the build-up in the combustion chambers reduces the clearance volume and so raises the compression ratio until, eventually, pinking will set in.

The valve seats, on the other hand, do not suffer from carboning. Instead, they become pitted with use, and in the end they no longer provide a proper seal. When that stage is reached—and for some hundreds of miles before it—the efficiency of the engine is steadily lowered. Pressure testing would show that the pressures within the cylinders were constantly being reduced as the valves became less and less able to close the ports fully. To restore this lost compression and to ensure that the full value of each working stroke is obtained is the real aim of decarbonizing.

Before the actual work begins make sure that you have all the equipment and material you will need. A torque wrench will be useful—even an improvised one if needs be; though many garages are prepared to hire out specialized items such as this. You will need a "decoke set" containing fresh gaskets for the cylinder head, the manifold, and the valve chest. A new set of valve springs should be obtained, too, for it is probable that the old ones will have lost some of their original resilience with use.

Besides the obvious spanners and so forth you will need in the way of tools a valve spring compressor and a large screwdriver for grinding-in the valves, a scraper to deal with the carbon, clean non-fluffy rag, paraffin or grease solvent, and valve-grinding pastes, both fine and coarse.

THE SIDE-VALVE ENGINE

First Steps in Decarbonizing. Make as much room for yourself in the driving compartment as you can, by removing the passenger's seat. Cover the floor pan with newspaper, then detach the engine cowling and the nearside engine shield to gain access to the power unit.

Open the bonnet, remove the radiator cap, and then open the tap at the bottom of the radiator and allow the coolant to drain away. If it contains anti-freeze which you wish to re-use you can drain it into a clean container.

While the water is draining you can start disconnecting the ancillaries. Working from the passenger compartment disconnect the throttle rod; detach the petrol pipe by undoing the union on the fuel pump; and then brush the unit with paraffin or grease solvent to remove all external oil and dirt. This is a wise precaution, since grit left on the engine at this stage may easily find its way inside and this could lead to serious damage later.

Manifold Removal. Turn, now, to the top of the engine. Detach the top chamber and the air bell of the carburettor. Unscrew (from below) the four bolts and brass nuts which secure the exhaust pipe to the manifold. Don't get these nuts mixed up with any of the others—it is vital that they should be replaced on this joint and no other, for their purpose is to save the fixing from rusting solid—which would happen were steel nuts used.

By undoing seven nuts you can now free the manifold completely. Slide it off its studs. If it jams, jar it very gently with a block of softwood, or with a fibre mallet.

Petrol Pump and Tappet Cover. Before you can reach the tappet cover the petrol pump will have to come off. Remove the second fuel lead—the one from the tank—and undo the two nuts which hold the pump to its studs. Draw it away carefully and do not lose any shims which you may find between the pump and the block. These shims are used to regulate the pressure developed by the pump.

Removal of two nuts now enables the tappet cover to be taken off. Here, again, if it tends to stick it may be jarred very gently with a wood block. *Never* try to lever it away by inserting a tool into the joint.

Head Removal. By now, the coolant will have drained from the engine and the next stage of dismantling can begin. Release the upper hose connection from the water outlet branch on the head. Disconnect all high-tension leads from the sparking plugs, and free the lead from the coil to the distributor and the lead to the battery. That done, take out the plugs.

Now scribe a mark from the distributor body to the clamping plate, and a second mark from the plate to the head. This will enable you to reset the timing easily when the time comes to rebuild the motor. Once you have done so, loosen the clamp's locking screw and pull the distributor from its housing, complete with its shaft extension.

Next, free the dynamo and remove the fan belt, the dynamo and its mounting brackets, and the water outlet branch.

Underneath an abutment at the front of the block, on the left side, you will find an upward-facing bolt. Remove this. This is vital, for if it is overlooked you may easily break the cylinder head as you try to lift it off. Free the upper cylinder head nuts and the head can be lifted. Once again, do not try to insert a sharp tool into the joint to lever it up if it does not free at once. This can result in a cracked head—or at best in damage to the joint faces. Instead, try jarring it with your block of softwood. If necessary, the head can be given quite sharp blows around the joint area with a hide-faced mallet.

If it is still recalcitrant, replace the sparking plugs and reconnect the battery. Then turn the engine over on the starter. The compression pressures should be sufficient to lift it and break the joint, but this procedure *does* mean that you will have to retime the unit later.

Work on Pistons and Head. With the head off bring two pistons to the top of their bores and then protect the other two by stuffing the cylinders with rag. Take the head to the bench, and use the scraper to remove all carbon deposits from the combustion areas. Pay particular attention to the area around the plug holes and pick away any carbon you find in the plug threads, using a needle. Finish off the job with a wire brush. If you use a rotary brush in an electric drill you will be able to give the surfaces quite a good polished finish which will resist subsequent carbon formation better than will a rough surface. Use the scraper, too, to remove any traces of old gasket adhering to the joint surfaces.

Earlier, I mentioned that carbon on the pistons can form a useful oil seal. Bear this in mind when dealing with a motor which has covered more than, say, 30,000 miles. On newer engines, it is permissible to scrape away all carbon from the piston crowns. But be careful, when doing so, not to allow the scraper to dig into the soft metal and so score the crown. Finish off by wire brushing till the metal is shiny.

On older engines follow the same procedure, but leave a ring of carbon, about a quarter of an inch wide, all the way round the edge of the piston. This is, obviously, a compromise procedure, but it is one which will

FIG. 17. PISTON CROWN DECARBONIZING
On old engines leave a quarter-inch ring of carbon round the piston to act as an oil seal. A sharpened stick of solder is a safe scraper.

ensure that you do not have to suffer a steep rise in oil consumption for a thousand or so miles after the job has been done.

When working on the pistons, deal first with the two which you have already brought to the top of their bores. Blow away all carbon dust, and then use the crankshaft pulley as a wheel and turn the engine over till the other two pistons are at t.d.c. Repeat the decarbonizing procedure for these, while protecting the other bores with clean rag. Again, blow away all carbon dust after the work is done.

Work on the Valves. Each valve is held to its seat by a spring, and this is retained by a cotter inserted through the valve stem. To remove the valves it is best to employ a special valve spring compressor. This is a form of G clamp which holds the valve onto its seating while the spring is compressed. The cotter pin is next pulled out from its place below the spring cap and the compressor is then released slowly, allowing the spring to expand. Take off the tool—when the valve can be lifted from the top of the block and the spring and cap taken out of the valve chest. On the head of the valve scratch a mark (Nos. 1 to 8) to show you to which seating it has to be fitted on reassembly.

When all the valves are out the first job is to clean them. You will find that the undersides of the heads—especially in the case of the exhaust valves—will have carbon deposits. Scrape these clear, and finish off with

FIG. 18. REMOVAL OF SIDE VALVES (1)
If you lack a compressor you can free the valves this way. Use two screwdrivers, as shown here, to compress the spring.

the wire brush. Also remove any carbon on the upper surfaces of the valve heads and make sure that the slots in the heads are clear.

Now comes the important grinding-in process. Having scraped and wire-brushed the areas round each valve port—and, of course, removed any carbon from the ports themselves—take the first valve and drop it back into its seating. If it will not seat fully, release the tappet adjuster and screw down the tappet until it will.

Now lift the valve and smear a little grinding compound on its face. Reseat it, and insert a screwdriver into the slot on the head. Rotate the valve back and forth through about 180 degrees half a dozen or so times. Then lift it, turn it through ninety degrees, seat it down again and repeat the process until on both the valve and the seat in the block there is one thin but continuous matt grey line. This gives a leak-free seating. Repeat the job with the other valves, until all are properly seated. To be sure about this you will have to wash away the compound with petrol before each examination. Purists also like to finish off each seat by a short grind with fine paste.

A word of warning though—examine each valve and seat carefully

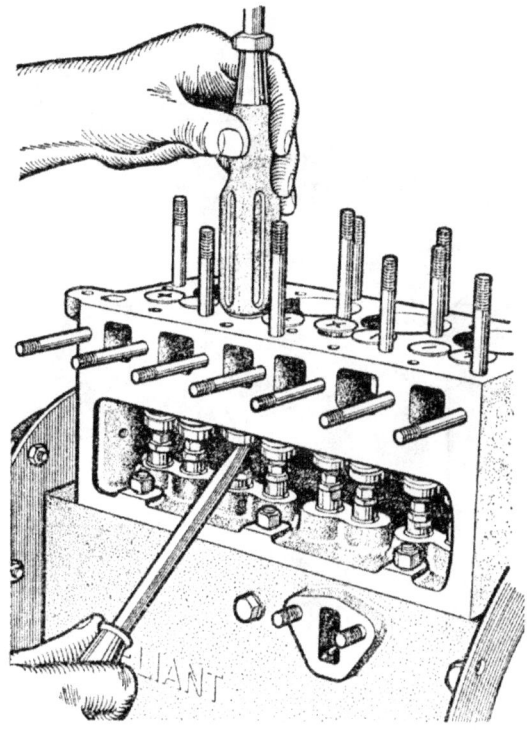

FIG. 19. REMOVAL OF SIDE VALVES (2)
Keep up the pressure with one screwdriver and tap the valve onto its seat with the handle of the other. This relieves tension on the pin.

before grinding commences. If there is really widespread and deep pitting on either the face or the seat it is doubtful whether grinding alone will give a satisfactory answer. A deeply pitted seat in the block will need refacing with a cutting tool—again, a garage may be prepared to hire you one—and a valve face which is badly affected will need the attentions of a refacing machine at a garage. In bad cases it may be better to fit a new valve. If you do, remember that this will have to be ground in just the same.

Valve Guides. After the grinding has been completed, and the valves and seats have been washed clean with petrol, turn your attention to the guides. Dip the valve stem into petrol or paraffin and insert it in the guide. Then lift it up and press it down, rotating it until all dirt is washed out of

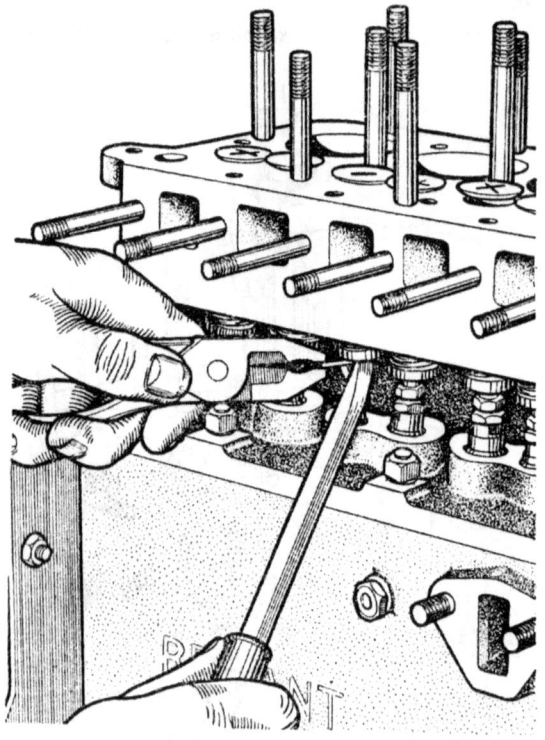

FIG. 20. REMOVAL OF SIDE VALVES (3)
Draw out the pin with pliers and then gently release pressure on the spring. The valve can then be lifted out of the block.

the guide. Remove the valve and wipe the stem clean. Then pass on to the next one, leaving the previous guide to dry out in the meantime.

When all have been cleaned, take the first valve and smear its stem with a thin layer of graphite grease. Drop it into its guide. Repeat with all the remaining valves.

Valve Spring Check. With use, valve springs gradually shorten and thus tend to lose efficiency. Check each valve spring in turn against one from the new set which you obtained. Some idea of relative condition can be gained by standing each used spring against a new one and placing a straight-edge over the top. If there is a gap between the top of the new spring and the top of the old with the ruler horizontal then the old spring should be discarded.

It is better, however, to make the check with both springs under pressure. This is done by interposing between them a suitable steel washer and butting their ends against it. The springs are then fixed between the jaws of a vice or clamp which is tightened until the new spring reaches its fitted length of 1·045 inch. Now measure the old spring. If it is more

FIG. 21. REMOVAL OF SIDE VALVES (4)
Alternatively, merely use this special compressor, fitted as shown here.

than ten per cent shorter than this, i.e. unless it is at least 0·94 inch long—it is better to discard it and fit the new spring instead.

Many owners believe—and I am inclined to agree with them—that for the sake of the few shillings involved it is best to fit new springs regardless of the conditions of the old ones. There is much to be said for this attitude, for a full decoke is carried out only at intervals of around 12,000 miles or so. As this may represent eighteen months or a couple of years' running the wisdom of "preventive" replacement of expendable parts such as these is obvious.

Refitting Valve Springs. Lift the valve until the spring and its cap can be slipped onto the stem. Insert the valve compressor tool, and compress the spring until the valve cap clears the cotter pin slot in the stem, when you can slip the pin into place. Slowly release the pressure on the spring, and the valve spring cup will descend onto the pin and lock it. Watch carefully, and if the cotter pin is not positioned so that the cup will fit squarely over it readjust it till it is. This is important, for if the cup and pin are off-centre the spring pressure may force the cotter pin out of its slot. Repeat this sequence with all eight valves.

Head Replacement. Make a final check to see that all traces of the old gasket have gone and that the studs for the head are clean. Take the new

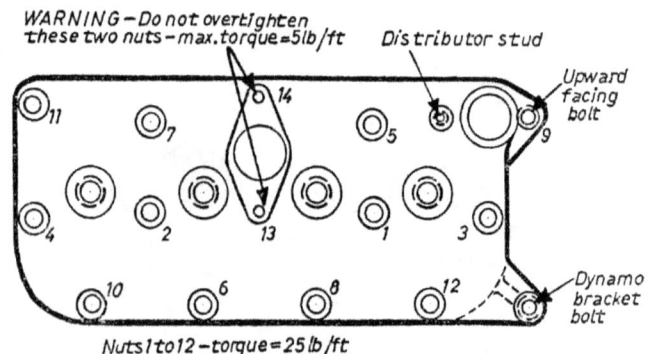

Fig. 22. SIDE VALVE HEAD REPLACEMENT
It is vital that the head nuts should be tightened in the order shown.

gasket and lightly smear both sides with jointing compound. Don't overdo this—a light coating is sufficient; too much may mean that compound will be squeezed out of the joint and into the mechanism.

Now fit the gasket, beaded edge downwards. Turn the crankshaft to bring No. 1 cylinder to T.D.C. on compression and then back until it is an eighth of an inch before T.D.C. Follow this by fitting the cylinder head. Seat it initially entirely by hand, ensuring that it is pressed evenly into place. You can now refit the ancillaries which are dealt with under separate headings.

Next comes the job of bolting it down, and here it is vital to follow the correct sequence of tightening the head nuts if distortion is to be avoided. As a start, replace all nuts finger tight. Then, using a torque wrench if possible, tighten each a few turns at a time—still working to the sequence shown here, until all are at the right tension. This job is done cold, first of all, and the settings are subsequently checked when the engine has been warmed up.

OVERHAULING THE ENGINE

Tappet Adjustment. When decarbonizing this job can best be done before the head is replaced, since it is then possible to see precisely when the valves are fully open. The procedure to follow is that set out on page 36 in Chapter 3.

Refitting Fuel Pump. After the tappet cover has been replaced the fuel pump can be connected up. Be certain to ensure that the pump's operating lever is inserted so that it lies over the camshaft. If it is inadvertently fitted *below* the shaft the pump will fail to operate and may be damaged. When you are sure about this tighten the two nuts fully.

Tappet Cover. It is vital that the cork sealing washer should be renewed at each major overhaul—and to guard against oil leaks it is no bad idea to use a new washer if the valve clearances are checked regularly at less than 2000-mile intervals. The action of tightening down the cover tends to compress the cork, and when it has been removed and replaced half a dozen times it is probable that it will tend to leak.

Bear in mind, also, that the cover must also be replaced with the line of breather holes uppermost.

Refitting the Distributor. Before fitting the head you pre-set the position of the crankshaft so that the timing would be correct one-eighth of an inch before T.D.C. on the compression stroke on No. 1 cylinder (engines 6001–15655) or at T.D.C. on compression, No. 1 cylinder (engines 19444 onwards).

The distributor is now inserted into its housing. Since the drive is offset it can be positioned only one way.

Realign the marks made before dismantling and tighten the clamp sufficiently to hold the distributor, yet to permit it to turn when twisted. Ensure by inspection that the contact-breaker points are just opening. If they are not, turn the body until this happens. A piece of cigarette paper slipped under the points will help you to discern the exact moment of opening. When this is found lock the clamp securely.

Final Assembly. Refit the manifold and carburettor, replace the dynamo and fan belt and the radiator top hose, reconnect the fuel pipes, replace the sparking plugs, and reconnect the electrical wiring. Connect up the throttle and refit the carburettor air bell. Finally, close the drain tap on the radiator and refill the cooling system.

Start the engine, and run it until it is warm. Then recheck the tension of the cylinder head nuts. Make a further check on all nuts and bolts after the car has covered 300 miles.

MAJOR ENGINE WORK

Major work can be defined as anything which entails removal of the cylinder block or the complete engine. Removal of the block would be required, for example, if the small end bearings were suspect, if the engine needed reringing, or for reboring and fitting new pistons.

Block Removal. Follow the same sequence of operations already described as far as removal of the cylinder head and valves. Then turn the crankshaft until all four pistons are at the same level. Remove the base nuts which hold the block to the crankcase, and lift the block carefully. Be particularly gentle as the piston rings begin to emerge from their bores, since it is easy to snap them.

Piston Removal. Remove the circlips which locate the gudgeon pin in the piston boss. Warm the piston by wrapping it in rag wrung out in hot water. This will expand the metal sufficiently to allow the pin to be pressed out. Mark the crown of each piston with its bore number, and with an arrow to indicate which face should be turned to the front when reassembling.

Reboring. This is specialized work which should be entrusted to an engineering workshop. It is usual for the firm which carries out the rebore to supply new pistons and rings of the correct oversized dimensions.

Oil Glaze. On engines which have seen many miles of use, an "oil glaze" tends to build up on the walls of the cylinders. This should be removed when new pistons or rings are fitted, other than at rebores.

The method—which is not as drastic as it sounds—is to employ a medium grade of wet-and-dry emery cloth. This must be thoroughly soaked in engine oil and then formed round a block to match the internal shape of the cylinder. This abrasive is then used to rub down the inside surfaces of the bore. By so "keying" the bore, one ensures that the new rings will bed down properly.

Wash the bores thoroughly with a generous amount of petrol after this operation.

Refitting Pistons. This is done as a reversal of the removal procedure, taking care to engage the gudgeon pin securing circlips into the slots in the piston boss.

Refitting the Block. Use of piston ring clamps $\frac{3}{8}$ inch deep is essential for replacing a Reliant block. And you will need to enlist the help of a friend.

OVERHAULING THE ENGINE

Fit the cylinder base joint first. Then place Nos. 2 and 3 pistons at the top of their stroke, and Nos. 1 and 4 at the bottom. Place the ring clamps round the two upper pistons and, holding the block perfectly horizontal, lower it onto these two pistons, using the clamps to feed the rings cleanly into the bore.

Now continue to lower the block until, with No. 1 and 4 pistons still at B.D.C., the top rings have entered and the mouth of the block is within $\frac{3}{8}$ inch of the bottom rings. Compress the bottom oil control rings and, very slowly, turn over the engine until these too have fully entered the cylinders. The clamps can then be withdrawn—making sure that they do not damage the base joint during removal—and the block fully lowered onto its studs. Tighten the nuts, and continue rebuilding as described in the decarbonizing routine.

Removal of the Engine Unit. Unlike many cars, in which engine removal calls for the availability of lifting tackle, the Reliant has been so designed that two men can remove the engine/gearbox assembly by hand.

First, disconnect the battery and drain the sump. Then remove the driver's and passenger's seats from the front of the passenger compartment and take out the carpets, engine cowling, gear lever knob, and the propeller shaft tunnel. This tunnel is held to the floor and heel boards by self-tapping screws. Remove the nearside footwell, and drain the cooling system. Disconnect the throttle rod and release the exhaust pipe nuts at the manifold.

Free the dynamo leads and detach the dynamo complete with the fan blades. It is held to the bracket by three bolts. Release the H.T. and L.T. leads from the distributor, and detach the oil pressure switch lead. Release the inlet and outlet hoses, sliding them from their respective pipes.

The clutch control mechanism must be freed. Remove the $\frac{5}{16}$ inch Oddie nuts and the springs and washers from the clutch rod, and loosen the bolt on the clutch lever clamp. This can then be tapped gently towards the back of the car till the clutch rod drops out.

Next the petrol pipe from the tank must be disconnected at the pump and the four five-sixteenths of an inch nuts and bolts on the gearbox coupling flange at the propeller shaft released. Disconnect the speedometer cable from the gearbox and release the large-diameter electrical cable from the starter motor. Take the earthing straps from the gearbox and remove the front and rear engine mountings.

Using a wooden block as a pad, place a jack beneath the sump and use it to raise the engine. When it is high enough, swing the tail end of the gearbox onto the nearside floorboard and ease the complete unit out of the car through the passenger's door.

Removal of Gearbox. Eight $\frac{5}{16}$-inch nuts, with spring washers, hold the gearbox to the engine. Remove these from their studs and slide the gearbox from the back of the engine. This exposes the clutch.

Removal of Clutch. To release the Borg and Beck clutch from the flywheel, merely remove the six set screws and washers around its periphery.

Stripping the Engine. Strip the bare engine as if for decarbonizing.

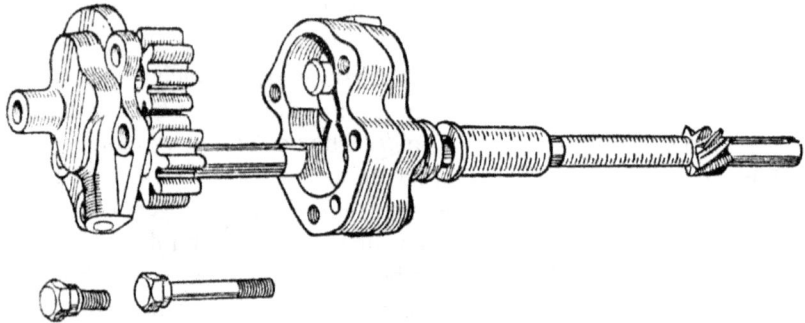

FIG. 23. OIL PUMP, SIDE VALVE ENGINES
All side-valve Reliants have this simple gear-type oil pump. It is driven from the camshaft by means of a skew gear.

Then detach the flywheel. This is held to its flange by three bolts, locked by a tab washer. On reassembly, a new washer will be essential.

Take off the rear engine plate, the crankshaft pulley, and the timing chain cover. Release the centre nut on the camshaft sprocket, and slide the timing chain and sprockets from the camshaft and crankshaft. Do not lose the associated Woodruff keys.

Next detach the front engine plate, the cylinder block (as already described), the oil sump and sump gauze, and the oil pump and pump gear spindle.

Working from the lower end, remove the connecting rods and pistons complete by freeing the two nuts securing each big-end cap. Mark each one so that it can be refitted in its correct position.

Take the two fixing bolts from the front camshaft bearing, remove the locking pin from the centre bearing, take the locking pin from the rear bearing, and withdraw the cover plate from the rear of the crankcase.

OVERHAULING THE ENGINE

Note that the camshaft centre bearing is in two halves, and that when withdrawing the shaft from the crankcase it is essential to so rotate it that No. 5 cam covers the split part of this bearing. It will then remove easily—just tap the rear end of the shaft, using a brass drift of half-inch diameter, until the centre bearing is free.

Finally comes removal of the crankshaft. Its front bearing is held by four bolts. Remove these, and tap the housing from the crankcase. Use a wooden drift to prevent damage.

From the centre bearing take the securing nuts and remove the bearing's lower half complete with bolts. On engines later than No. 8267 remove the locking wire from the centre main bolts and unscrew the bolts from the centre bearing *and* the bolts holding the bearing cap into the crankcase. In both instances the top half of the bearing is removed by use of a small brass drift tapped lightly round the crankshaft. It is vital that this should be done carefully, or damage to the light alloy crankcase may result.

Now detach the four bolts which hold the rear main-bearing housing to the crankcase. Then—to enable the two Allen screws which join the two halves of the bearing to be removed—push the crankshaft backwards in the case. With these out the housing can be parted and the crankshaft lifted from the crankcase.

Refitting the Camshaft. Stand the crankcase vertically on its rear end and insert the camshaft until No. 5 cam is in the centre bearing housing. Then insert the bearing, keeping it square to the housing all the time, with the split half horizontal. The shaft can then be lowered on to the top of the centre bearing. Use No. 4 cam to cover the split half of the bearing while the shaft is tapped, lightly, to drive the bearing home. Alternate the position of No. 4 cam from the inner half of the bearing to the outer as you do so. When it is home, drive in the locking pin.

Using jointing compound on the joint washer, fit the rear camshaft bush, its locking pin, and the rear cover case. Then add the front bush, and secure it with its twin locking bolts.

Test for end float by fitting the sprocket without its key and tightening it down. If the gap between the sprocket and front bearing exceeds 0·008 inch renew the bearing.

Refitting the Crankshaft. Before the crankshaft can be correctly positioned to refit the front and centre bearings the rear housing and bearing must be installed on it. Place the jointing washer over the shaft, split the rear bearing housing, and ensure that both housing faces are absolutely clean. Then smear the front faces only with a light coating of jointing compound. Install the housing around the shaft and tighten the two Allen screws. If a torque wrench is available set them to a tension of 180/200 pound inches.

It is vital that housings should enter the crankcase squarely. To ensure that they do, "borrow" a couple of the exhaust manifold studs from the cylinder block and screw them lightly into the case to act as guide dowels.

Lift the crankshaft, keeping it horizontal. Pull it forward with the left hand, at the same time using your right hand to lead the rear bearing housing squarely onto the stud until it enters the crankcase. Then use a hide mallet to tap the back of the housing until it is firmly seated. You may now take out the two studs and insert the four set-screws with spring washers, which secure the housing to its seat.

Next, the front bearing must be fitted. Place the white metal thrust washer to the rear of the front bearing housing and offer up the whole unit to the nose of the shaft. Slide it into the crankcase, and insert and tighten its four fixing bolts. These, too, have spring washers.

Now slide on the white metal thrust washer and follow this with the steel thrust washer. Slip on the crankshaft sprocket and pulley—at this stage without the Woodruff keys—and tighten them up so that the amount of thrust can be measured. The limits are 0·010 inch minimum to 0·020 inch maximum. If there is more than this new thrust washers must be used.

At this stage the centre bearing can be replaced. When bolting it on ensure that the castings number on the bearing housing faces the front of the engine.

To fit the bearing merely reverse the procedure already detailed for bearing removal. Where locking wire is used, renew it. *Never* take a chance on using the original wire a second time.

Once all the main bearings have been reinstalled, take a set of feeler gauges and measure the clearance between the crankshaft oil thrower and the rear bearing housing. Take measurements at various points round the thrower. None should exceed 0·003 inch, and the thrower must be squarely set.

When replacing the centre main bearing on engines after No. 8267 inclusive—in which there are two ⅜-inch studs fitted to the centre web of the crankcase—first slide half of the bearing housing with the bearing shell into place in the case and then fit the other half to the two studs. Tighten down the two securing bolts and lock them with copper wire. *Never* use wire which has already been employed on this job. It may fracture and lead to severe internal damage. Finish the job by adding the two ⅜-inch Oddie nuts and tightening them fully.

OVERHAULING THE ENGINE 57

One other word of warning—when fitting the top half of the bearing it is essential to ensure that the oil feed groove to the camshaft centre bearing does in fact lead in the right direction.

Final Assembly. Now reinstall the block. With the engine out it is possible to do this single-handed. Fit the pistons back onto their connecting rods, without bearing caps, and offer them up into their respective bores. Push them halfway in. Note that while the offset on the small

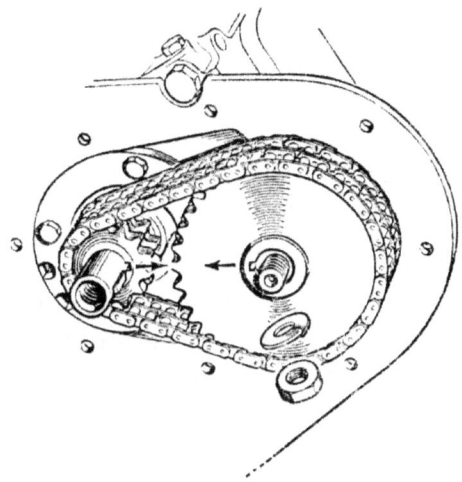

FIG. 24. RETIMING SIDE VALVE ENGINES
To reset the timing, as described in this chapter, the Woodruff keys on the crankshaft and the camshaft must be brought into line.

end of No. 1 rod must face to the front, those of the other three rods point to the rear of the block.

Install the cylinder base gasket and turn the crankshaft so that all the big-end journals are horizontal. Then lower the block onto its studs and fasten the securing nuts.

That done, rotate the crankshaft to bring Nos. 1 and 4 big-end journals to the top and ease the respective con-rods onto the journals, fitting the bearing caps loosely. Remember that new nuts and bolts should be used. Then bring Nos. 2 and 3 journals to the top, and similarly connect their rods. The big-end bolts may now be tightened to a torque of 250 pound inches.

Remove the crankshaft and camshaft sprockets. Fit the oil pump assembly and the front engine plate. Then assemble the crankshaft and camshaft sprockets with the timing chain. Rotate the crankshaft so that the keyway points direct to the camshaft. Then turn the camshaft so that its keyway is pointing to the crankshaft. Position both Woodruff keys. Similarly align the keyways on the two sprockets. Then, keeping the sprockets aligned, slide the sprockets back onto their shafts, tapping them until they are both fully home. Replace the spring washer and nut on the camshaft and tighten it, and then fit a new timing case gasket followed by the timing case. That done, the crankshaft pulley can be fitted and bolted down.

The remainder of the reassembly work is a straightforward reversal of the procedures already outlined for stripping.

THE OVERHEAD-VALVE ENGINES

For decarbonizing assemble the requisite tools and spares: a valve spring compressor, socket spanners, a suction-type valve-grinding tool, wire brushes, feeler gauges and cleaning material, and a complete decoke set including new gaskets, valve stem oil seals, and a spare set of valve springs.

Dismantling. Drain the cooling system. Besides the radiator drain cock there is a drain plug for the cylinder block—a hexagon-headed set screw placed in the block beside the breather pipe on the left side of the unit.

Disconnect the battery and remove the H.T. leads from the sparking plugs. Then take out the plugs themselves.

Detach the choke cable, the accelerator linkage, the pipe from the fuel pump to the carburettor, and the suction pipe leading from the base of the carburettor to the distributor.

Disconnect the top water hose as soon as the coolant has drained, then turn to the exhaust pipe. Release this at the manifold joint and press it down as far as it will go.

At this stage thoroughly clean the exterior of the unit, for you are about to open up the internal mechanism and it is vital that no dirt should find its way inside. You can brush it with paraffin; or else use a grease solvent washed off with water. In that case, dry away the water with absorbent rag before proceeding.

Next, take off the rocker cover. Working from above, remove the nuts on the rocker shaft pedestals and lift the complete rocker gear off its studs. Have ready a piece of card punched with eight holes. Number these from front to rear. Twist each pushrod several times to break any suction between the base of the rod and its tappet block, and then lift each rod in turn and place it in its numbered hole in the card. It is a good idea to have holes, also, to receive the valves when they are removed.

The cylinder head nuts may now be freed and the head lifted. If it ends to stick jar it with a block of soft wood or with a hide-faced mallet.

OVERHAULING THE ENGINE

Decarbonizing. In general, follow the procedure for decarbonizing described on pages 44–49, with the obvious reservation that the valve seats and the ports are in this case in the head and not in the cylinder block.

Remember, also, that in the 3/25 engine you are dealing with light alloy combustion chamber surfaces, and that you must take extra care not to damage the relatively soft metal. For this reason, it is best to use as a scraper a stick of hard solder filed at one end to form a wedge shape.

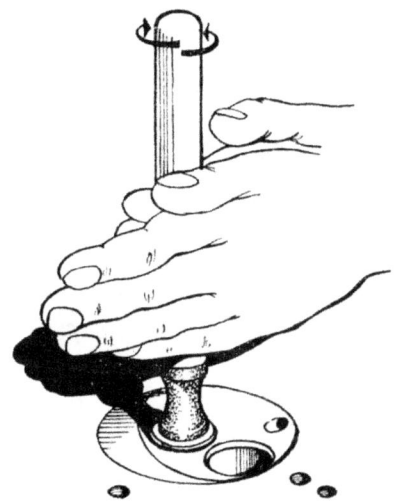

FIG. 25. GRINDING-IN THE VALVES
A suction tool, as shown here, must be used with the overhead valves. Those on the side-valve engines are slotted to take a screwdriver.

This will deal satisfactorily with the carbon, but will not be hard enough to harm the alloy.

If you plan to use wire brushes in an electric drill take care not to allow the steel centre of the brush or the drill chuck to come into contact with the edges of the combustion chambers, or they may easily be indented. And, of course, it is vital to keep the face of the cylinder head free from damage, or water and gas leakage may result.

Many mechanics, when wire-brushing an overhead-valve engine's head, prefer to reinstall the valves temporarily so that the valve seats are safeguarded, too. This practice is to be recommended. What should *never* be done, however, is to use emery cloth for cleaning away carbon. Emery is highly abrasive, and any dust particles trapped in the engine will cause very heavy wear.

Valve Replacement. After grinding, use the valve spring compressor to refit the valves, having first checked the springs against new ones as advised on page 49.

Insert the valve and, holding it onto its seat, slide the oil sealing ring over it. Fit the spring and retainer, slide the tool into place, and compress the spring. Then insert the collets, using a dab of grease to hold the first in position while you insert the second into its groove in the valve stem. Gently release the compressor. This allows the retainer to rise and trap

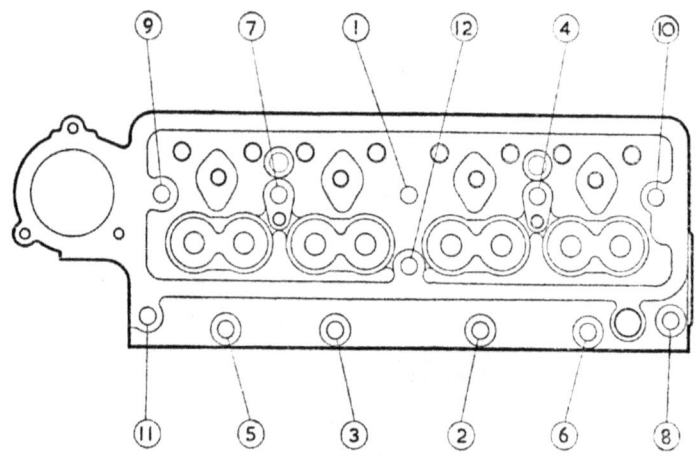

FIG. 26. O.H.V. HEAD REPLACEMENT

Whenever the head nuts are tightened down this sequence must be followed, or there is a very real danger that the head may be badly distorted.

the collets. Make sure these seat squarely. Remove the tool, place the head firmly on the bench, and with the end of the handle of a hammer press the valve down once, releasing it smartly. This checks that the collets have in fact seated properly.

Reassembly. After cleaning all jointing faces smear clean engine oil around the cylinder bores. Rub a light coating of grease—*not* jointing compound—onto the new cylinder head gasket and place it over the studs. Lower the head onto the block and, following Fig. 26, tighten the head nuts to a tension of 25 pound feet.

Now replace the pushrods, refit the rocker shaft, and tighten its pedestal nuts. Adjust the tappets as detailed on page 37, and then rebuild the rest of the unit by reversing the dismantling procedure.

OVERHAULING THE ENGINE

Overhaul. This is not a difficult unit to strip. Where an overhaul is thought desirable the following methods should be adopted. Removal of the engine/gearbox unit from the car is advisable.

Removing Engine/Gearbox Unit. Drain the radiator, disconnect the fuel pipe at the carburettor, and the hoses from the radiator and heater. Free the H.T. lead from the coil to the distributor and detach the starter motor lead.

Next detach the throttle and choke controls, the exhaust down pipe, and the inlet manifold, complete with carburettor. Remove the exhaust manifold, the dynamo, and the starter motor.

Next detach the gearbox cover—complete with gear lever—and jack up the front of the car. Block it thoroughly so that it is safe to work underneath, and then free the front end of the prop shaft, followed by the clutch operating rod. This is detached at the cross-shaft end.

At the point where the engine stay rod passes through the rear engine plate (on the right side of the engine when viewed from the rear) free the stay. Disconnect the speedo drive—at the lower rear right side of the gearbox—and the leads to the oil warning light switch and the water temperature gauge.

Now unscrew the engine mounting bolt nuts at the front end, place a jack under the gearbox (using a block of wood as a "pad" between the jack and the gearbox casing), and undo the rear mounting bolt's nut. When this bolt is removed the engine/gearbox unit can be lowered to the ground, and the vehicle lifted away from it. Where a hoist is available it is advisable to use it to ease the lowering process.

Detaching the Gearbox. Eight $\frac{5}{16}$ inch nuts hold the gearbox to the rear engine plate. Remove them, and ease the gearbox away from the engine.

Detaching the Clutch Unit. The clutch unit is held to the flywheel by half a dozen $\frac{5}{16}$ inch set screws. Release these by loosening each in turn a few threads at a time, so that the unit comes away squarely.

Dismantling the Engine. On the right side of the unit you will find the fuel pump. Remove the pipe connecting the pump to the carburettor—the carb. itself will already have been detached—and the $\frac{7}{16}$ inch by $\frac{3}{4}$ inch A.F. bolt which secures the distributor. This can then be removed, complete with its leads and vacuum advance/retard suction pipe.

Remove the sparking plugs. Then detach the fuel pump—held by a pair of $\frac{1}{2} \times \frac{3}{4}$-inch A.F. nuts. Do not lose the lock washers or the pump gaskets. Next unscrew the oil filter and pressure switch, and remove the dipstick.

Turning now to the left side of the unit, unscrew the ten $\frac{1}{2}$-inch A.F. brass nuts holding the inlet and exhaust manifolds—if these were not, in fact, detached before removal of the unit—and disconnect the head-to-pump by-pass hose in the cooling system. Take off the dynamo by unscrewing one $\frac{5}{16}$-inch A.F. and two $\frac{1}{2}$-inch A.F. nuts and bolts. When detaching this unit, be careful not to lose the spacer which fits between the dynamo and the front engine plate.

Now begin work on the top of the unit. Unscrew the rocker cover nuts and detach the cover. Slacken the four rocker support bracket nuts ($\frac{9}{16}$ inch A.F.). To guard against distortion release each nut just a few threads at a time, so that the rocker assembly remains square to the head. When the nuts are off lift away the rocker assembly complete, and twist out the eight push rods.

All is now clear for lifting the head. Again the twelve $\frac{9}{16}$-inch A.F. cylinder head nuts must be released evenly, each loosened a few threads at a time. Break the head joint by jarring with the heel of the hand—do *not* try to lever or hammer the head to break the joint—and lift the head gently. Finally, take out the tappets. Be careful to keep these in their proper order.

To strip the bottom half of the unit, set out wood blocks on the workbench so that the engine can be turned onto its left-hand side and yet remain steady. Then remove all the sump bolts—thirteen in number on early models; fifteen on later cars. They are $\frac{7}{16} \times \frac{3}{4}$-inch A.F., all supplied with lock washers. The spacer, sump, and gasket can then be detached.

Two $\frac{1}{2}$-inch A.F. nuts retain the oil pump, which should be lifted out complete with its housing and filter.

Next the eight big-end locking plates can be relieved and the $\frac{7}{16}$-inch A.F. big-end cap bolts unscrewed. Take off only one cap at a time, press the piston and connecting rod out through the top of the bore, then loosely refit the cap to its rod. This will ensure that these important components remain matched.

If the clutch has been left untouched to this stage the crankshaft must now be prevented from turning by interposing a block of hardwood between the crank and the case while the peripheral bolts are unscrewed. Then, with the clutch detached, free the flywheel locking plate and take out the three $\frac{9}{16}$-inch A.F. bolts which secure the flywheel to the crankshaft. The flywheel, complete with ring gear, can be removed by the use of two levers inserted between the flywheel and the case.

OVERHAULING THE ENGINE

The rear engine plate is held by three $\frac{1}{2} \times \frac{3}{4}$-inch A.F. bolts and by one 1-inch A.F. bolt. After this has been taken off, remove the crankshaft thrower cover. This has six $\frac{7}{16}$-inch A.F. bolts. When doing this part of the job be very careful not to damage the sealing ring.

Next turn your attention to the front of the engine. With the crankshaft still held by the hardwood block, take out the pulley-retaining bolt and its washer, and remove the pulley. Take off the timing cover (held by eight $\frac{7}{16}$-inch A.F. bolts) to bare the timing chain and sprockets.

The camshaft sprocket has a locking plate, whose tabs must be flattened. After removal of the two $\frac{1}{2}$-inch A.F. securing bolts this sprocket will pull off, bringing with it the timing chain.

A puller will be needed to remove the crankshaft sprocket. Do not lose the Woodruff key which locates it on the shaft—and take similar care not to mislay the steel thrust and Vandervell washers. Then detach a further pair of $\frac{1}{2}$-inch A.F. bolts and a retaining plate.

Take off the two $\frac{7}{16}$-inch A.F. bolts which hold the front engine plate, detach the plate, release the timing chain sprayer (held by one nut) and the four cheese-head screws and bridge pieces at front and rear. After detaching the three main bearing caps—held by two nuts apiece and marked "Front," "Centre," and "Rear" for easy identification—the crankshaft and its remaining thrust washer can be lifted from the crankcase. Take off the oil sealing ring.

On the oil pump extension shaft is a $\frac{1}{8}$-inch Mills pin. Remove it and, with a suitable drift, drive the shaft through the distributor bore. This frees the camshaft.

Normally, no further dismantling will be necessary; but it is possible, at this stage, to remove the cylinder liners for replacement. This is done by tapping them gently from inside the block, near the bottom of the liner, and lifting the loosened liner from the top. New O-rings *must* be used when the liners are refitted.

The camshaft bearings can be driven out after removing the core plug set in the rear of the block. When fitting new bearings make sure that the oil feed hole is in line with the oil gallery. If this is not done severe oil starvation and rapid bearing failure will result. To complete stripping the oil pump spindle bushes and the oil relief valve and spring can be removed.

Reassembly of the unit involves nothing more complicated than a reversal of the stripping procedure, but it is essential to use a torque wrench for all nuts and bolts and to follow the torque settings given in the tables in the Appendix. This will avoid the danger of stripping vulnerable light alloy threads, or of cracking or distorting the material.

COOLING SYSTEMS

All Reliants are water cooled, but whereas the side-valve engines have a simple thermo-syphon system the overhead-valve models use a pump for water circulation.

In both cases, however, the basic care of the cooling system is the same. Before the warm weather sets in it is advisable to check the cooling system thoroughly. Where anti-freeze has been used—unless, of course, it is one of the "permanent" varieties—the system must be drained and thoroughly flushed with a hosepipe to remove all impurities.

So far as the radiator is concerned it is as well, once a year, to employ reverse flushing. First drain the system. Then remove the bottom hose, and rig up a garden hose so that water can be forced in through the lower orifice.

Obtain a length of old cycle inner tubing and stretch this over the radiator's filler orifice to form a drain-away conduit. Then turn on the tap, forcing water through the radiator from bottom to top. This flow, which is in the reverse direction to normal, will dislodge deposits which would not be touched by flushing in the normal direction.

Once a year, too, it is advisable to remove the radiator from the car and direct a hose jet straight through the honeycomb. Dirt and dead insects trapped in the air passages can have the effect of reducing the cooling area, and hosing is the only way to deal with such obstructions.

Examine all hoses for perishing at least once every year. Ensure that the clips are tight. Also examine the radiator cap. As the cooling system is pressurized each cap has two valves. The larger spring-loaded valve is designed to lift from its seating if internal pressure becomes excessive. Mounted concentrically is a smaller valve connected to the outside atmosphere. This allows air to enter if pressure in the system falls below atmospheric pressure. Were this to jam it would be possible for air pressure to crush the radiator.

Most garages have pressure-test equipment which will tell you instantly if your cap is operating correctly. If you are at all unsure about this, it is best to have such a check made. At least, satisfy yourself that the relief valve opens when sucked lightly, and that the valves are both seating properly.

Every winter anti-freeze should be added to the water to prevent damage to the block and radiator. If the water in the cylinder jackets freezes the pressure set up as the water turns into ice, and consequently expands, is enough to crack the metal of the block.

Because of the danger of corrosion only the recommended anti-freeze solutions listed in the Appendix should be used in the all-alloy engine, *save in an emergency.* Where the choice is between temporary use of the wrong anti-freeze and leaving the unit unprotected *any* anti-freeze can be used for a period of up to a week or so without harm. But if this expedient has to be adopted it is vital that the whole system should be thoroughly flushed when the stop-gap solution is drained and before the correct anti-freeze is added.

7 The Fuel System and Carburettor

ON both side-valve and overhead-valve models the fuel system of the Reliant is quite conventional. At the rear of the car is a cylindrical fuel tank, into which is inserted a fuel pipe and filter. Also in the tank is the sender element for the fuel gauge—a form of rheostat operated by an arm and float.

Fuel is induced into the pipe by means of a pump mounted on the engine. Driven by a cam on the camshaft this pump feeds fuel from the tank to the carburettor, pump pressure being so set that it is insufficient to overcome the resistance of the carburettor's needle valve assembly when the float chamber is full. Generally speaking, the result is a simple and utterly reliable layout which will normally call for the very minimum of care and maintenance.

To say that, however, is not to imply that it requires *no* care and maintenance. It does, for a number of good reasons. One of these is, of course, that petrol delivered from a garage pump is not a clinically pure liquid, no matter how carefully it is handled. All pumps are fed from huge underground tanks, and there is invariably a small water content in the fuel. Most of this sinks to the bottom of the car's fuel tank—which is why the tank pick-up is so located that it does not draw out the last dregs of petrol—but some inevitably gets carried into the rest of the system.

All fuel also contains small solid impurities—one reason why filters are incorporated. Again, though the filters catch most of this dirt a proportion does in fact reach the carburettor, and after a considerable mileage—15,000 miles say—some at least of the jets will have become scored by the passage of these minute pieces of grit. Hence, the carburettor will no longer give of its best; while further down the system the pump and tank filters will steadily be growing restricted as dirt builds up in them. And at the bottom of the tank the trapped water may eventually set up corrosion if air, too, can reach the metal—as it may, for example, if the fuel level is allowed to fall so low that the tank is virtually empty. Hence, cleaning the fuel system periodically must be regarded as an essential item of maintenance. And even more essential is overhaul of the carburettor at intervals of around 15,000 miles. For this purpose the carburettor manufacturers—Solex Ltd., Solex Works, 223-231 Marylebone Road, London, N.W.1—supply a special set of parts known as an Econokit. This comprises all the more vulnerable components—gaskets, jets, etc.—which experience has shown will be in need of replacement by such a mileage.

CARBURETTOR OVERHAUL

With an Econokit, the carburettor can be given a new lease of life in an afternoon—always providing, of course, that the instrument is not so old that wear around the butterfly has become excessive.

Remove the carburettor from the engine and thoroughly clean the outside in a bath of petrol. Allow it to drain, then wipe it with rag so that no dirt remains and the surface is dry.

Spread clean newspaper on the bench and remove the top of the float chamber. Lift out the float. Next, take out the jets and the emulsion tube. As you remove each jet make a note of exactly where it was originally placed. This will avoid the unpleasant possibility of having a stripped carburettor and a series of jets which might fit anywhere.

Take out the volume control screw, together with its spring. Then swill out the float chamber in clean petrol to remove all sediment. Give the various channels a wash, too, and blow through them with a tyre pump. Don't forget to operate the pump half a dozen times, to rid it of any dirt, before applying it to the carburettor.

You can now start rebuilding. Open the Econokit and select the new pilot jet, main jet, and washer for the main jet carrier bolt. Fit these, pilot jet first. Next, replace the starter petrol jet, using the washer provided in the kit. Add the new volume screw. That done, refit all remaining jets.

Fit the float back in its chamber making sure, with 28 ZIC-2 carburettors, that the toggle is the right way up. Then replace the float chamber lid, using the new gasket provided.

Test the carburettor flange with a straight-edge to ensure that it is not bowed or warped. If it is not absolutely true it should be faced-up by grinding-in with coarse paste on a sheet of plate glass. If you are quite satisfied with the flange clean from the mating surfaces of the carburettor and the manifold any remnants of the original washer and replace the instrument on the new washer provided in the kit.

You can now reconnect the pipes and linkages and restart the engine. Test the various joints for leaks. You can best test the carburettor flange joint by "painting" petrol around it with a brush. If there is a leak some of this petrol will be drawn into the intake and the engine will speed up slightly.

It should be noted that carburettor gaskets should always be fitted dry. Never use jointing compound, grease, or oil. The reason is simple. Under pressure any of these substances will be squeezed out, and may block one of the fine internal passages.

Econokits do not include new throttle spindles or butterflies, since no serious wear should have taken place in these units within the 15,000 mile period. Where the carburettor is older than this, however, it is advisable to fit new parts here too. Release the butterfly plate from its spindle by undoing the screws which hold it, and take the control arm, etc., from the spindle. The whole throttle is then removed. If there is no undue wear in the carburettor casting a new throttle assembly will suffice. Where the metal in the body of the instrument has become ridged, however, it is better to hand in the old instrument for an exchange unit.

After fitting a new carburettor, or reconditioning the old one, it is always advisable to check the contact-breaker points and the plugs so that the improved efficiency of the carburation is not offset by an out-of-tune ignition system.

SOLEX 26 AIC AND FAI CARBURETTORS

Setting the Idling. With the engine hot set the throttle adjustment screw so the engine is idling fast. Then slacken off the volume control screw till the unit begins to "hunt"—i.e., the running becomes irregular. At this point reverse the action of the volume control screw, and continue to screw it in until the hunting just ceases. Then slow the engine down by releasing the throttle adjustment screw. If this causes more "hunting" turn the volume screw slowly clockwise until the engine again idles evenly. The ideal speed is about 500 r.p.m.

Access to Jets. In this carburettor the pilot jet, main jet, starter air jet, and starter petrol jet are all easily accessible from the exterior of the instrument.

The pilot jet is located under a small plug high on the side of the instrument, and the main jet under a larger plug fitted lower down. The starter petrol jet is under a blind plug immediately beneath the bi-starter unit, and the air jet is on top of the starter.

When jets are removed for cleaning they should be cleared by blasting air through them in the reverse direction to the normal fuel flow. Serious obstructions can be cleared by poking through with a bristle, but it is vital never to use wire or the point of a needle or pin to clear obstructions. Jets are made of soft metal, and there is considerable danger of the accurately-drilled metering holes being enlarged if metal is used in them.

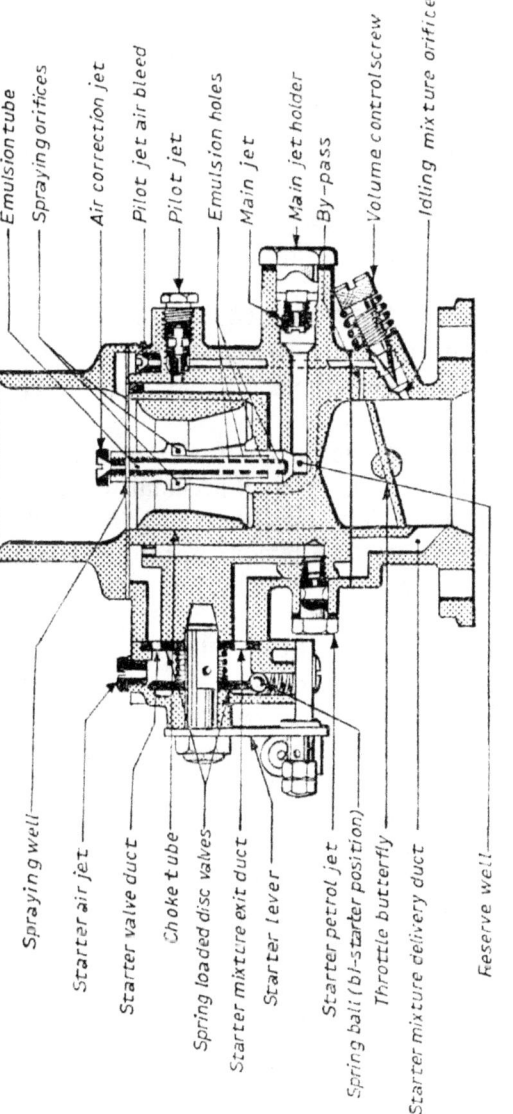

FIG. 27. SOLEX 26 AIC AND FAI CARBURETTOR

This cross-sectional view clearly shows the disposition of the various major parts of the instrument.

For much the same reason—the possibility of damage to the carefully-contoured surface—the volume control screw must never be tightened hard against its seating.

SOLEX B 28 ZIC-2 CARBURETTOR

In general, the instructions set out here for the Model 26 AIC and FAI carburettors hold good for the B 28 ZIC-2. However, there is a difference in the siting of the starter jets. In this instrument the starter petrol jet is placed behind a blind plug on the float chamber, while the starter air jet is interposed in the float chamber air vent system and can be reached only when the float chamber lid is lifted.

AIR CLEANERS, ALL MODELS

Cleaning. With use, the oil-wetted wire mesh of the air cleaner becomes choked with dirt. If this is not cleaned the engine will run rich and waste fuel.

Detach the air cleaner and thoroughly wash the gauze in clean petrol, using a stiff-bristled brush to remove all accumulated dirt. Then allow the cleaner to drain dry. Next, soak the gauze in oil. Leave it fully immersed for fifteen minutes, then remove it from the oil and allow the excess to drain away. Wipe the filter body, and refit it to the car.

THE A.C. FUEL PUMP

Every 2500 miles the filter must be examined, and cleaned if needs be. It is located below the top cover, which is held by a single control screw. Remove this, and the filter can be lifted from its seating and cleaned by swilling it in petrol. At the same time flush out any sediment which is trapped in the pump chamber.

Before refitting the cover it is vital to ensure that the rubber washer is in good condition. If it is damaged it is better not to attempt to make do but to fit an entirely new one. Remember, also, to refit the fibre washer under the head of the retaining screw, and to tighten the screw itself enough to make a petrol-proof joint. But don't overdo it, for the screw is threaded into soft metal and if you apply too much force you will undoubtedly strip the thread.

Stripping the Pump. Release the petrol pipes at both inlet and outlet unions and remove the two nuts holding the pump to the crankcase. The pump can then be lifted away.

Make a vertical mark across the joint between the top and bottom castings with a file. This will aid reassembly. Then wash the pump

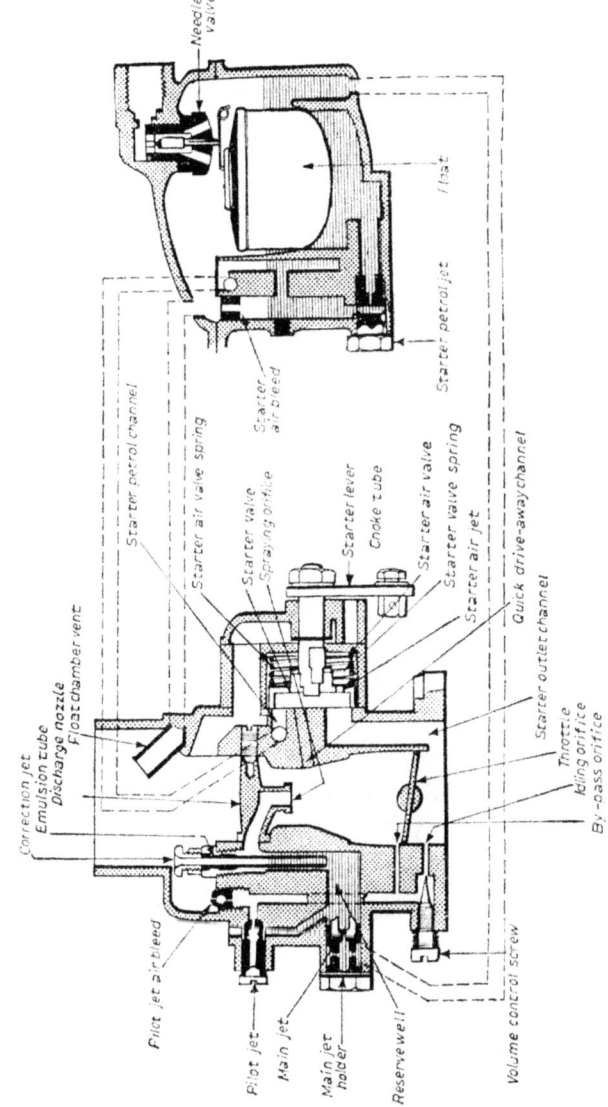

FIG. 28. SOLEX B 28 ZIC-2 CARBURETTOR

Although of similar design this unit, used on the o.h.v. 3/25 model, has differences of detail in its layout.

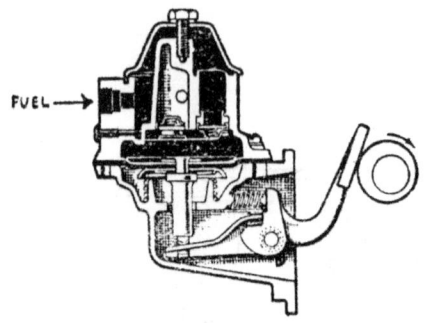

SUCTION STROKE

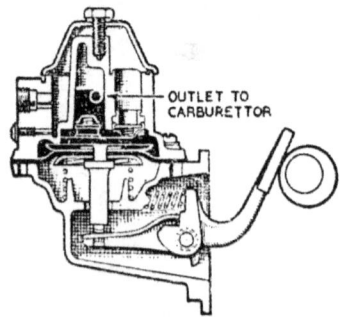

DELIVERY STROKE

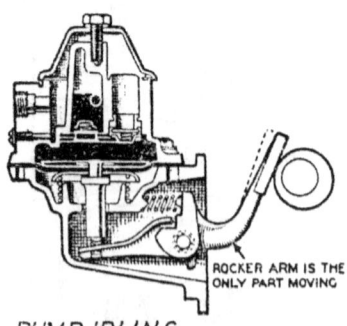

PUMP IDLING

FIG. 29. THE A.C. FUEL PUMP
Three phases in the operation of the A.C. mechanical fuel pump, which is driven by an eccentric on the camshaft.

THE FUEL SYSTEM AND CARBURETTOR

thoroughly in clean petrol before dismantling begins. Remove the screws which hold the two castings together. The lower half will contain the diaphragm mechanism. The diaphragm and pull rod can be detached by turning it through ninety degrees and lifting it away from the body.

It is usual to practise preventive replacement of the pump's diaphragm assembly whenever the unit is dismantled. In any case, it should be noted that where the top of the pull rod is riveted to the top washer no attempt should be made to separate the diaphragm layers from the protective washers and the pull rod. All gaskets should also be renewed.

Wash the area around the valves in the upper casting in a paraffin bath and examine both halves for cracks or any other damage. Check the filter cover, and if it proves to be distorted inwards around the centre hole renew it. A badly corroded filter gauze should also be renewed, as should the diaphragm spring. These springs are colour-coded for strength, and it is vital that the colour of the original (or the number stamped on the pump flange) should be quoted when obtaining a replacement.

Examine the valves, springs, and seats for deterioration and the link engagement slots for wear. Worn rocker pin holes mean that the pin must be peened to give a tighter fit. Invariably, the rocker arm spring should be replaced, and wear at the point where the rocker arm makes contact with the link must not be greater than to permit a slight slackness when fitted to the rocker arm pin.

Fuel Pump Reassembly. On YD type pumps, where the valves, springs, and seats are held in the upper casting by a retaining plate and screws, ensure that the "8" shaped gasket is in place before the assemblies are fitted. On the inlet valve the spring must be fitted protruding into the chamber, while the outlet valve is fitted in the reverse position so that it will allow a flow of fuel out of the pump. This done, fit the retaining plate and tighten its two screws.

Assemble the link, packing washers, rocker arm, and rocker pin in the lower half of the body. Fitting of the rocker pin can be simplified by first inserting a piece of 0·24-inch diameter rod through the pin hole on one side of the body. Press it in far enough to engage the rocker arm washers and link. Then enter the rocker arm pin from the other side, easing out the temporary pivot as the pin takes up position.

Where the pump has an oil seal, fit this and its spring onto the pull rod and turn the washer through ninety degrees to hold it in place. Position the diaphragm assembly over the spring, pull rod pointing downwards, and centre the upper end of the spring in the lower protector washer. As you press down on the diaphragm turn the assembly to the left to set the pushrod in its correct working position relative to the link. This will take a quarter of a turn. It will enable the pull rod slots to engage with the fork on the link, and will also align the holes in the diaphragm with those on the pump body. Note that when first placing the diaphragm assembly in the body the locating tab will be in the eleven o'clock position.

You can now assemble the pump. Push the rocker arm towards the pump until the diaphragm is level with the body flanges, and then place

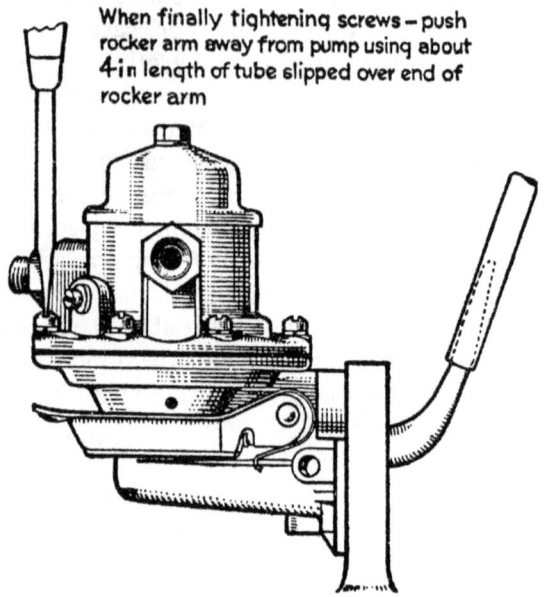

FIG. 30. FUEL PUMP RE-ASSEMBLY
To ensure proper operation of the pump this procedure must be followed whenever the unit is rebuilt after the body halves have been parted.

the upper half of the pump on top, aligning the marks made before stripping.

Fit the screws and lock washers and tighten just sufficiently for the screw heads to engage on the washers. Now slip a four-inch length of tube over the end of the rocker and press the arm away from the pump. This will hold the diaphragm at the top of the stroke. Keep up the pressure so that the diaphragm remains there while the cover screws are fully tightened, working from one screw to that diagonally opposite.

Pumps which are fitted with rocker arm stop screws and washers must have these parts removed during the operation just described. When the two halves are securely mated the stops can be refitted.

Note that when matching-up is correct the diaphragm edge should be more or less flush with the two clamping faces. If there is any appreciable protrusion the job must be done again, and special care taken to keep the downward pressure on the arm during final tightening of the screws.

Pump Troubles. There are about half a dozen possible troubles which may occur with the fuel pump. Fuel starvation at high speeds can be attributed mainly to general wear and tear in an old pump, or to air leaks caused by loose unions or damaged or over-compressed cork filter gaskets.

Fairly closely associated with this is difficult starting caused by slow priming. It can often be due to leakage or wear, faulty valves, or a diaphragm which is insufficiently flexed. But it is also possible that the fault may lie in the carburettor float chamber. If this leaks the fuel level is reduced and the pump must first supply enough fuel to restore the correct level before a start can be made.

Pump pressure is controlled, in the first instance, by the diaphragm spring. Thus carburettor flooding which sets in when the pump has not been disturbed indicates that the fault lies in the float chamber. However, in such a case—if it is persistent—the delivery pressure can be adjusted to help matters by fitting one or two extra gaskets of normal thickness at the pump/crankcase joint. Obviously this packing out must not be carried to excess, or high-speed starvation will result.

Excessive wear on the pump itself can normally be due only to poor lubrication, most likely on old engines, where blow-by past the pistons causes corrosion in the lower half generally. The obvious solution here, of course, is to overhaul the engine.

8 Work on the Electrics

PERHAPS at first sight one may be forgiven for wondering what on earth there can be in an electrical system to maintain. But second thoughts will show that there is a very considerable amount of mechanism involved. Not all the electrics, by any means, consists of static wiring.

Three essential parts of the car's electrical equipment are, in fact, basically mechanical: the distributor, the dynamo, and the starter motor. So, too, is the windscreen wiper mechanism. All these components require periodic attention, and after varying mileages will also need a thorough overhaul.

Besides these, there is the battery—the only chemical part of the layout. Now, a Reliant-sized battery cannot be overhauled—though the big accumulators used on buses, for instance, are in fact rebuilt several times during their working lives. But it still needs to be checked and maintained. If it is not it is likely to fail no less certainly than the engine would fail, for example, were its lubrication to be neglected.

Even the wiring, static though it may be, cannot be wholly overlooked. Deterioration can set in—either through age or because the harness has rubbed against a sharp metal edge—and to guard against a possible "burn out" it is advisable to make a close visual check on it from time to time.

THE DISTRIBUTOR

General Maintenance. The need to check the contact-breaker points setting at regular intervals was mentioned in Chapter 5, where the method of making adjustments is fully described.

Periodically a lubrication schedule should be carried out. When doing so be very careful not to get oil or grease onto the contact breaker points, and do not overdo the oiling, or lubricant may be thrown onto the points as the distributor operates. Again, the details of the lubrication routine can be found in Chapter 5.

Cleaning is advised at 5000-mile intervals. Use a soft dry cloth to wipe both inside and outside the distributor cap, paying particular attention to the space between the internal terminals.

Test the H.T. pick-up brush in the centre of the cap. This is spring-loaded into its holder, and when it is pressed it should move freely. If it sticks, or feels stiff, remove it and find out why. There may be some dirt between the brush and the walls of the holder, or it could be that the spring has been damaged and is no longer doing its job properly.

Now take off the rotor arm, and examine the contact-breaker points. If they are badly burned and pitted it is best to fit a replacement set.

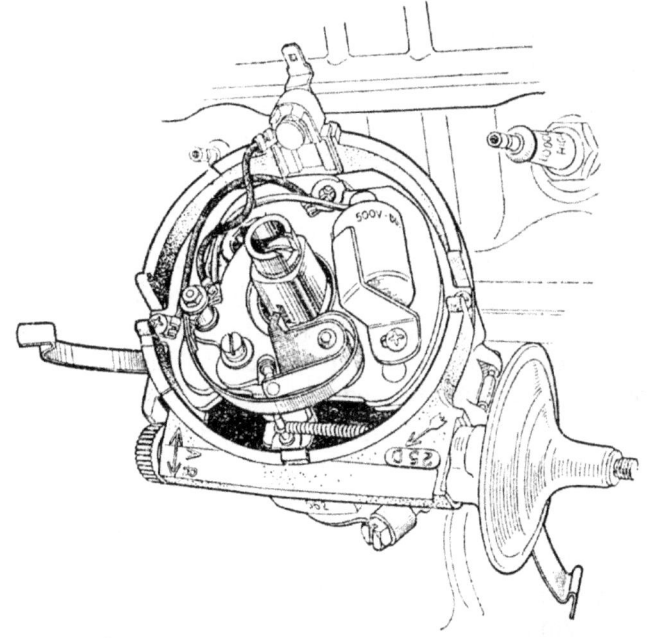

FIG. 31. LUCAS DISTRIBUTOR, O.H.V. ENGINE
Unlike the distributor on earlier Reliants this type incorporates a vacuum advance/retard unit, whose operating spring can be seen attached to the base plate.

Moderate surface deterioration can be rectified either by stoning on an oil stone at home or, better still, on a garage's points refacing machine. In either case, wipe the points clean of metallic dust and grease afterwards by using non-fluffy rag moistened in clean petrol.

Before replacing the points wipe the pivot post clean with petrol-moistened rag. Then smear it lightly with Mobilgrease No. 2 or with Rogosine Molybdenized non-slip oil. Reinstall the points; set them as detailed in Chapter 5, and replace the rotor arm and the distributor cap.

Distributor Overhaul. After a very considerable mileage has been covered—around the 40,000-mile mark—the distributor will be ripe for overhaul. This is a basically simple procedure, and one which is quite within the ability of the average owner.

Remove the distributor from the engine, as described in Chapter 6, and detach the distributor cap and the rotor arm. Next, free the drive shaft dog. This is locked in place by a small locking pin inserted through the driving dog. Use a punch to tap this out, holding the body of the distributor in such a position that the smaller section of the offset is on the left of the body, with the shaft in such a position that the rotor arm would be pointing upwards.

Free the screws holding the contact-breaker base plate, and detach from its pillar on the plate the spring from the vacuum advance/retard mechanism. The entire base-plate assembly, including the contact points, can then be drawn out of the distributor body. Side-valve models do not have a vacuum-controlled advance/retard mechanism, relying entirely upon the centrifugal automatic advance/retard unit which is incorporated inside the body. On these cars, therefore, it is now possible to withdraw the distributor shaft complete with centrifugal weights and springs. On the overhead-valve engines the vacuum unit must first be dismantled. Do this by removing the small circlip which is fitted onto the shaft just in front of the vernier adjuster knob. A pair of pliers must be used here —it is likely that the circlip, which is fairly brittle, will break. Consequently, it is advisable to have a spare circlip handy.

Unscrew the vernier knob and then draw the vacuum diaphragm unit away from the distributor body. That done, the shaft and bobweights can be withdrawn. Stripping the centrifugal mechanism is a straightforward operation. Simply ease the advance weight springs off their pillars on the cam plate. But be very careful not to distort them in doing so, or new ones will certainly be required upon rebuilding.

Fitting a New Bush. Almost undoubtedly it will be necessary, with an old unit, to renew the distributor bush. Drive out the old one—taking very great care not to fracture the body material. The replacement bush should have been soaked in oil for twenty-four hours before fitting. This gives it virtually life-long lubrication, though obviously it still requires small quantities of oil to be injected from time to time.

Use a stepped drift to drive in the new bush. Then, using the existing oil hole in the body as a guide, drill a lubrication hole in the bush. Offer up the shaft. If it is an easy fit, and turns without binding, all is well. If not, have it reamed at a garage to give the desired clearance.

Shaft Check. In the top of the shaft is a central set screw, securing the cam. Release this, lift the cam, and remove the centrifugal weights. The bare shaft can then be checked for wear using, preferably, a micrometer. It is possible, however, to employ a vernier gauge quite successfully. If there is obvious ovality it is best to fit a new shaft. In other cases check the area around the hole for the drive dog pin. Burring frequently occurs here. If this has happened, file away any metal which is standing proud.

Distributor Reassembly. Clean the centrifugal weights and their associated springs and plates in petrol. Allow them to drain and dry thoroughly, and then lubricate them with a molybdenum grease. Then reassemble the advance/retard mechanism to the shaft and refit the cam.

Place the shaft/cam assembly back into the distributor body. In the case of overhead-valve engined cars, rebuild the vacuum advance/retard mechanism. First, offer up the diaphragm unit. Then, from the other side, slide on the ratchet plate and coil spring. Add the vernier knob, screwing it on until it is in the halfway position. Fit the new circlip.

Unless the existing contact-breaker points are in particularly good condition it is advisable to use a brand-new set. Fit them to the base plate. Be sure that you don't get your fibre washers mixed up when doing this. Refit the leads.

Now lightly coat the cam with grease or the recommended oil (see page 77) and offer up the complete base-plate to the body. Where a vacuum unit is fitted remember to hook its spring onto the pillar as you do this. Reconnect the earth lead, and gap the contact-breaker points. This is very easily done at this stage. Just turn the shaft until the heel of the fibre contact-breaker arm is riding on the peak of the cam; then set the gap to 0·015 inch and replace the rotor arm.

Finally, fit the driving dog. Turn the rotor arm so that it lines up with the low tension terminal, the contact plate on the arm pointing towards it. Position the dog so that the smaller off-set is on the left side and drive home the locking pin. Reinstall and time the distributor, following the instructions given under the appropriate heading in Chapter 6.

THE DYNAMO

Lubrication. At 12,000-mile intervals on side-valve engines (or once every year, whichever is the more frequent) a few drops of clean S.A.E. 30 engine oil must be injected through the hole marked "Oil" which you will find at the commutator end of the instrument. That apart, the dynamo requires no lubrication at all between overhauls.

On overhead-valve engines a different type of dynamo is used, and this requires no attention at all between overhauls, since both bearings are prepacked with grease.

Brushgear Checking. Remove the dynamo from the car. To do this, disconnect the *D* and *F* terminals and detach the fan belt. Unscrew the three bolts which hold the dynamo to its bracket on the head.

With the dynamo on the bench detach the driving pulley and remove the terminals from "output" and "field," then unscrew the two through bolts which hold the dynamo end plates. As these bolts are withdrawn

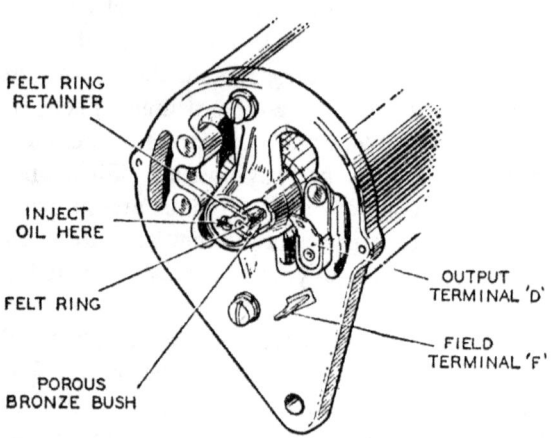

FIG. 32. DYNAMO LUBRICATION
A felt ring in the commutator end bracket gives wick-type lubrication for the bush
Oil it regularly.

remove the brush housing. Then the driving end bracket and the armature assembly can be detached from the yoke.

Test the brushes by lifting each into its brush box and hold it there by slipping the spring onto the side of the brush. Fit the commutator end bracket over the commutator, and release the brushes.

Hold back each brush spring in turn and gently move the brush thus freed, tugging very gently on its flexible connector. It should move freely. If it doesn't, remove the brush completely from the holder and lightly ease its sides with a smooth file.

Where brushes are obviously badly worn they must be renewed and the replacements properly bedded down. To do this, wrap medium-grade glass paper round the commutator, abrasive side upwards, and press the working surface of the brush onto it while the commutator is revolved. This grinds away the carbon of the brush until its contour exactly matches that of the commutator. On the Reliant the minimum permissible length of the brush is $\frac{9}{32}$ inch.

Armature Recutting. In time the process of wear and tear on the armature will result in the metal segments no longer standing sufficiently

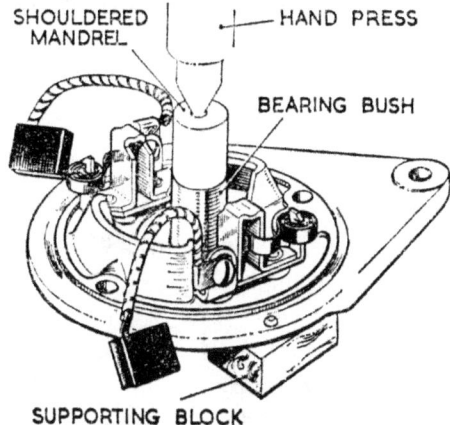

Fig. 33. Bearing Renewal on a Dynamo

A press is used to install a new bush on a dynamo end bracket. Note the carbon brushes and the springs which retain them in their housings.

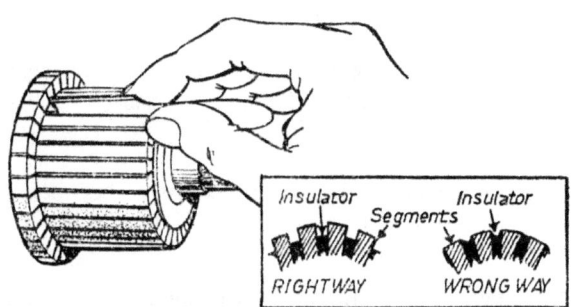

Fig. 34. Cutting Back Insulation

The insulation between the segments of the commutator must be kept cut back as shown here. When this becomes necessary use a broken section of hacksaw blade for the job.

proud from the insulation. When this stage is reached the insulated sections must be cut back. A section of hacksaw blade broken off squarely will be found a suitable tool for this. It is essential that the job be done cleanly and that the cuts should follow the actual shape of the segments —a wedge.

Dynamo Testing. The armature of a dynamo may be tested by use of a moving-coil voltmeter, or even by means of a wander lead attached to a bulb of the same nominal voltage as the dynamo itself.

With the unit on the car remove the dynamo leads and connect the voltmeter or leads to the D terminal and to earth. Set the engine to run at a fast tick-over (no faster, or the dynamo may be damaged) and in the

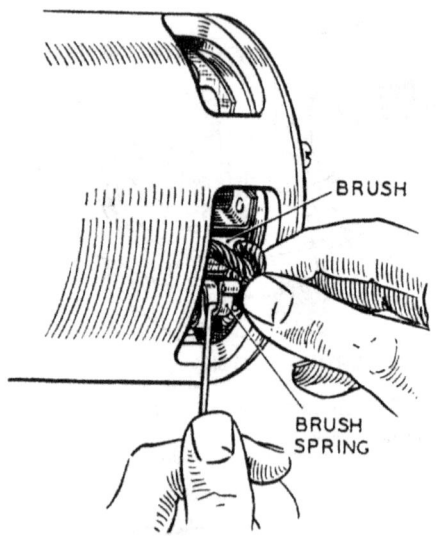

Fig. 35. Brush Removal Through "Windows"
Where the dynamo has "windows" the brushes can be hooked from their housings without stripping the generator.

former case, the voltmeter should give a reading of between 2 and 4 volts. If a bulb is used it should light.

Refitting Brushgear. To remove the brushgear the end plate will have been taken off the dynamo. Take this opportunity, therefore, of cleaning the bearings in petrol, allowing them to drain dry, and then repacking with high melting point grease.

Where the dynamo has "windows" cut in the yoke the brushes can be reinstalled after the end plate has been offered up, and it is merely necessary to ensure that when this is done they are fully home in the housings and that the springs are properly reset.

"Windowless" dynamos, on the other hand, call for a different method. Before replacing the end plate lift each brush and trap it with the spring as for checking. Offer up the plate with the brushes raised and refit the two through bolts. Then insert through the end slots a length of steel rod of a diameter small enough to enter the rolled end of the spring, and release the pressure of the spring on each brush in turn. The brush will

FITTING C.E. BRACKET TO "WINDOWLESS" YOKE GENERATOR

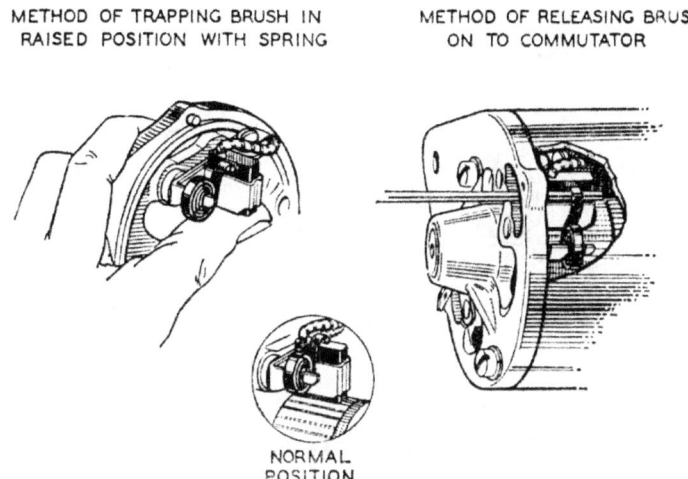

FIG. 36. REFITTING AN END BRACKET

On "windowless" units the end bracket has to be removed to reach the brushes. Refit them by using a piece of steel rod, as shown. The operation is detailed in this chapter.

slide down its holder, and the spring can then be released to take up its normal position on the top of the brush.

THE STARTER MOTOR

Testing in Position. Where starter motor trouble is encountered there are several possibilities which can be tested very easily with the unit still in position, assuming that the battery is fully charged.

Switch on the lights and press the starter button, listening carefully. If the lights dim but you cannot hear the starter operating it indicates that the current is flowing through the starter's armature windings but that the armature is not turning. If the engine is not abnormally stiff— as it would be, for instance, following a seizure—the probability is that the starter pinion has become permanently meshed with the ring gear. This could happen as a result of the starter being operated while the engine was running.

If, on the other hand, the lamps retain full brilliance when the starter button is operated either the supply circuit is broken or else there is an internal fault in the motor itself.

Where the starter action is very sluggish there is a possibility that the fault lies in a loose connection, with consequent high resistance in the circuit. A fast starter action, which does not crank the engine, points directly to damage to the drive itself.

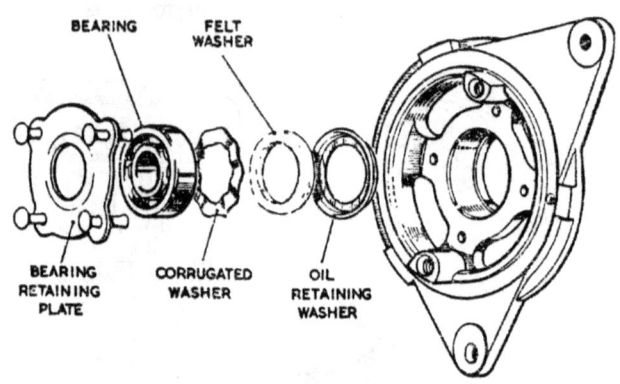

Fig. 37. A Dynamo Rear Bearing Stripped
A riveted plate conceals the roller-type rear bearing which is a drive fit in its housing.

Removal of the Starter Motor. Troubles or no, every 15,000 miles the starter must be removed from the engine. Before attempting to do this DISCONNECT THE BATTERY. Failure to do so may result in a short circuit and can lead to a very nasty burn.

Next, disconnect the main lead from the starter motor and take out the two bolts which hold it to the engine plate and clutch housing. Pull the starter motor clear. Should it stick it can be jarred loose by gentle tapping with a wooden block.

Stripping the Starter Motor. The first step is to dismantle the drive. Remove the split pin which locks the shaft nut. Then place a spanner on the squared end of the shaft—in front of the commutator—and with the shaft thus locked undo the shaft nut. They will come off in the following order—main spring, buffer washer, corrugated retaining ring (inside the pinion barrel unit), control nut, sleeve, and restraining spring. If the splined washer on the armature shaft is now removed the pinion barrel itself can be slipped out of place.

Next, detach the metal band cover from the commutator end of the body ("yoke") and lift the brush retaining springs. Holding them back, lift the brushes from their holders.

Take the nuts off the terminal post which protrudes from the commutator end bracket and unscrew the two through bolts. The bracket can now be detached from the yoke. That done, the driving end bracket will come away—and the armature will come with it. You may find a thrust washer on the commutator end of the shaft. If so, be careful not to lose it.

Commutator Cleaning. If it is to operate efficiently the commutator must be clean, and not fouled by oil, grease, or dirt. Loose dirt can be

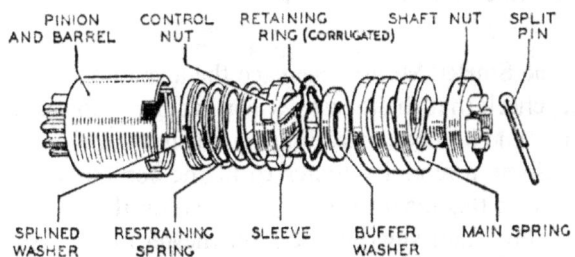

Fig. 38. Starter Motor Pinion Unit
An exploded view showing the quick-threaded sleeve.

removed with a soft clean cloth, more stubborn dirt with cloth dipped in petrol. Where the contamination is particularly bad, wrap a piece of fine glass cloth—*not* emery cloth—around the commutator and around this wind a strap formed from a length of stout cloth. With the commutator held firmly in a padded vice, pull the ends of the strap up and down. This will rotate the glass cloth and thus polish the commutator surface.

Unlike the dynamo commutator, the insulation between the segments of a starter motor must *never* be undercut.

Brush Renewal. Two of the brushes on the starter motor are attached to the brush boxes on the commutator end bracket, and two are connected to the aluminium field coils.

Brush wear is very limited, but if any of the brushes has reached the stage at which it no longer bears on the commutator, or if wear is so pronounced that sections of the flexible connection are exposed on the running face, renewal will be necessary.

It may well be better, in such a case, to entrust the work to a qualified auto-electrician. In case of emergency, however, it *can* be done at home if great care is taken with the soldering. This is the method to adopt.

One-eighth of an inch (3 mm) away from the aluminium, cut cleanly through the flexible connection. Clean up and tin the original resistance-brazed joint. Then take your replacement brush unit, and open up the loop in its flexible connection. Tin this—but, in doing so, be absolutely certain that you do not allow any solder to run towards the brush.

Offer up the new brush unit to the original connection, placing this joint within the loop. Then squeeze up the loop, and make a strong soldered joint.

Repeat the sequence for any other brushes which need to be replaced. There will be no need to bed the new brushes to the commutator, since all spares are preformed during manufacture.

Rebuilding the Starter Motor. Replace the armature/end cap assembly into the yoke, check that the thrust washer has been refitted and then add the commutator end bracket, threading the brushes through the "windows" in the yoke. Some mechanics prefer to fit the commutator end bracket first, believing that this makes it easier to position the brushes.

Replace the fibre and steel washers on the terminal post, fit the nut, and replace and tighten the two through bolts. That done, slip the brushes back into their holders and secure them with the springs. Replace the cover plate.

The drive can now be rebuilt. Fit the pinion barrel first, followed by the splined washer, the restraining spring, the sleeve, the corrugated retaining ring, buffer washer, main spring, and shaft nut. Lock the nut with a new split pin. It should be noted that if the control nut or the screwed sleeve happens to be damaged *both* components must be replaced, not just the individual item.

THE BATTERY

Singularly little in the way of care and maintenance is required by the battery, but neglect will drastically reduce its life. During use, evaporation takes place within the battery cells. What is lost is not the working acid, but part of the water content. It is this which must be made good if the battery is to continue to function.

In hot weather check the electrolyte level weekly. During winter, fortnightly checks will suffice. The correct level is just above the tops of the battery plates, which can be seen when the plugs are removed.

Where the level has fallen, add *distilled* water. On no account should tap water be used, for it contains impurities which might cut battery life. Where no distilled water is available, an emergency substitute can

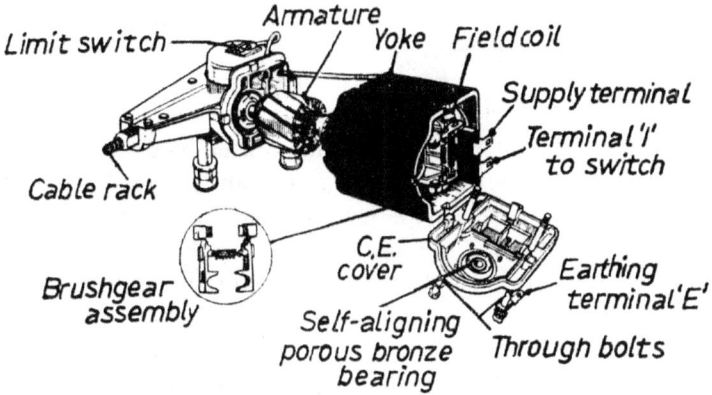

FIG. 39. WINDSCREEN WIPER MOTOR
Of simple construction the wiper motor may none the less need occasional attention to its brushes and commutator.

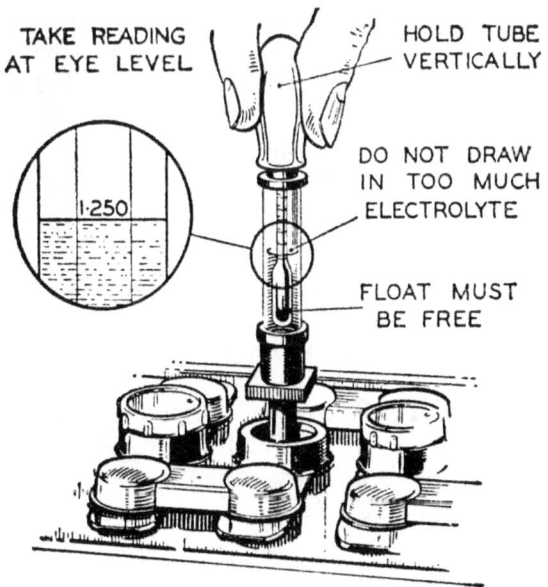

FIG. 40. HYDROMETER TESTING A BATTERY
Checking the specific gravity of the electrolyte with a hydrometer is the only way to determine its condition. A reading of 1·250 indicates proper charging.

be obtained by using the water which results from defrosting the freezer unit of a domestic refrigerator.

It is equally important not to overfill the battery cells. If this is done electrolyte will be forced out of the cells and damage the surrounding metal. Where the battery has been inadvertently overfilled you can remove the surplus quite easily. Obtain a clean length of plastic piping with an internal bore of around a quarter of an inch. Cut off a length of some six inches and insert one end into the overfilled cell. Place your finger over the top hole, keeping it there as you lift the tube out, and a quantity of electrolyte will be trapped in the tube. Pour this away, and repeat until the level is correct.

Do not allow corrosion to build up around the battery terminals. Periodically, remove the battery from the car and clean the terminals and their posts. Neutralize the acid on the terminals by immersing them in a mildly alkaline solution (soda bicarbonate is suitable) and finally washing them in clean water and drying them thoroughly.

The posts should be wiped with an alkaline solution to remove and neutralize any corrosion. Then polish each post with fine emery cloth until it is shining. That done, coat it with Vaseline. Replace the battery and refit the terminals. As they are tightened they will displace some of the jelly and make a good electrical connection. After tightening is complete, apply Vaseline to the terminals also.

From time to time, check the specific gravity of the electrolyte with a hydrometer. Take the reading at eye level. It should be 1·250. A battery in a very low state of charge is indicated by readings of the region of 1·190–1·120. This may indicate that the dynamo or control system is out of order.

Readings above 1·250, coupled with a need for over-frequent topping up and, possibly, burned-out bulbs are evidence of the contrary state of affairs—a system in which the battery is being consistently overcharged. If the charging rate with all lights and accessories in use and the car running is over one amp it is best to take the car to an auto-electrician and have the regulator unit checked.

9 Clutch, Gearbox, and Transmission

THERE are four main components comprised in the Reliant transmission system: the clutch, gearbox, prop shaft, and rear axle. However, the amount of practical work which can be done on these is limited, and most private owners would probably prefer to leave well alone so far as the gearbox and rear axle are concerned.

Obviously, also, it is impossible to work on such units as the clutch, gearbox, or differential without a considerable amount of dismantling, and unless ample under-cover facilities are available it is not even advisable to attempt to do so.

Gearbox Removal. It is essential to remove the gearbox for work on itself or on the clutch. Follow the procedure set out for engine removal in Chapter 6, with the difference that the eight bellhousing nuts are freed beforehand so that only the gearbox is in fact lifted from the car. The clutch is now accessible. On the 3/25 the car is blocked up, and the rear end of the box is dropped on a jack to remove it from below.

Clutch Removal. Six $\frac{5}{16}$-inch bolts secure the clutch unit to the flywheel. Undo each of these a little at a time, so that the assembly comes evenly away from the flywheel.

Should the flywheel itself also need to be taken off, straighten out the tabs on the lock washer securing its fixing bolts. Release the three bolts, using a $\frac{5}{16}$-inch spanner, and lift off the flywheel.

Clutch Renovation. A private owner will be well advised not to attempt any form of renovation, but to use a service exchange unit instead. This is a ready-prepared cover assembly and new driven plate which is simply substituted for the original unit. The only special tool then needed is a mandrel with which to line up the driven plate boss with the flywheel spigot bush.

Unless the flywheel face has been damaged there will have been no need to remove it from the engine; but if the flywheel has been detached a new tab washer *must* be employed to lock its bolts. Old tab washers must never be reused, since there is a danger of the weakened tabs fracturing and permitting the securing bolts to loosen.

The mandrel is inserted into the flywheel bush and the new driven plate is fed over it, followed by the cover assembly. This is bolted down securely, and the spigot is then removed. Once this has been done it is essential not to operate the release mechanism before the gearbox has been refitted, or the driven plate will be decentralized and the clutch will have to be removed again to permit realignment.

A mandrel can be improvised from an old main shaft purchased from a breaker's yard, or even from the handle of a suction-type valve grinding tool suitably shaped to fit the bush and the splines.

After rebuilding the engine/gearbox unit and installing the ancillaries readjust the clutch operating rod to give the correct play (one to one and a half inches free movement) at the clutch pedal.

Dismantling the Gearbox. As a general rule the oil should be drained from the gearbox before removal. Do this by removing both filler and drain plugs. Then remove the gearbox as described. Wash the exterior with paraffin or grease solvent so that no danger exists of external dirt entering the unit during stripping.

Lever out the two retaining springs from the clutch housing end of the box and withdraw the clutch release bearing and cup assembly. If this is badly worn it should be discarded in favour of a new bearing.

Unscrew the set-pin in the collar on the left side of the box and turn the unit so that the bell housing is horizontal. In this position the split taper pins which hold the operating fork can be removed. With a brass or steel drift, with a minimum diameter of half an inch, tap the pins until they are level with or just below the operating fork. Then use a taper punch to drive the pegs through the operating fork and cross shaft. This may now be withdrawn.

Detach the eight screw and washers which hold the top cover; free the rear coupling set screws and main shaft nut locking washer. Then, with a screwdriver, operate the gear selector to engage one of the forward gears and also reverse gear at the same time. This locks the main shaft, so that the nut which secures the rear coupling to the shaft can easily be undone.

Remove the rear cover and press out the main shaft, using a hide mallet or a hammer with a block of softwood interposed. Tap the shaft lightly from the rear cover, the ball race and speedometer worm gear also being detached at the same time.

From the main shaft, press off the ball race and speedometer gear.

CLUTCH, GEARBOX, AND TRANSMISSION

Then detach the front ball race and oil seal housing by undoing the four quarter-inch set screws which hold the front housing, and driving them out with the hide mallet.

The reverse operating lever is held by a pivot pin. This passes through a bush, which is itself positioned in the side of the gearbox by a pair of ¼-inch set screws. To detach the lever, unscrew the self-locking nut which secures the first pin in the bush and tap the pin through into the gear

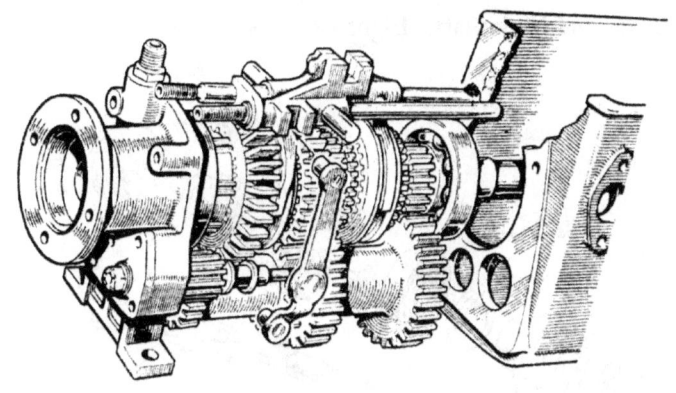

FIG. 41. THE GEAR CLUSTER

This is the basic gear cluster, with the selectors at the top. The gear lever engages in the slots to operate the selector forks.

nut. The bush itself need not be removed, and in general it is not necessary to detach the reverse gear shaft mounting during gearbox overhauls.

Before the primary gear can be removed the lay gear must be dropped into the bottom of the box. A self-locking nut holds the lay shaft. Remove it, and tap the shaft from the housing. This is secured by a pair of ¼-inch bolts.

Remove the shaft through the rear end of the gearbox, letting the lay gear fall into the gearbox well. Then use a tyre lever to trap the ball race in the front housing and tap out the primary gear with a hide mallet. Keep tapping until the gear is clear of the ball race—and if the race itself should protrude from its housing tap it back into position so that the lay gear can be removed.

Thrust Washer. Take particular care of the hardened thrust washer which is located between the end of the main shaft and the bottom of the lay gear. This determines the amount of thrust between the primary gear and the main shaft.

Selector Shafts. These are removed by, first, ensuring that the gears are in neutral and then unscrewing the two locking nuts at the rear of the gearbox. The shafts are then unscrewed through the bell housing. When the threaded portions of the shafts are clear of the rear of the gearbox use a length of rod (maximum diameter, ¼ inch) to tap them out. Be careful not to lose the spring and ball in the selector forks, which can now be lifted out of the unit.

Rear Cover and Mainshaft. Eight fixing bolts—two short and six long

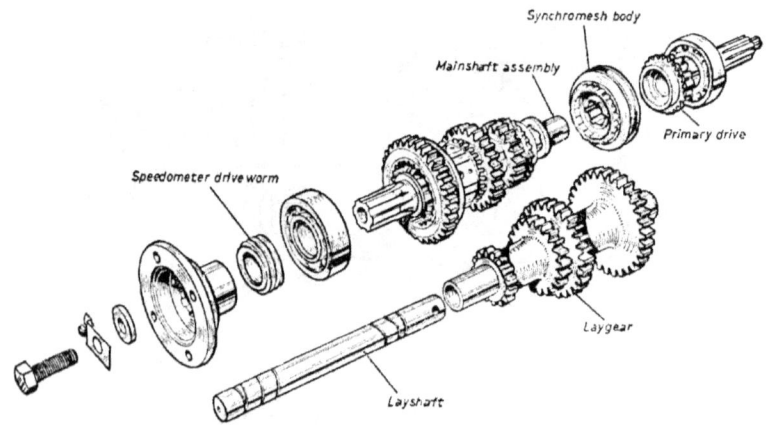

FIG. 42. THE GEAR SHAFTS "EXPLODED"
This drawing clearly shows how the two shafts are built up and how the rear flange is fitted.

ones—hold the rear cover to the gearbox. Detach these and remove the speedometer bush and pinion and the rear cover.

To obviate any danger of the feet being broken off the cover it is advisable to take off the housing complete with main shaft. Use a brass drift to tap the main shaft lightly, on the raised portion of the splines, until the rear housing is free of the box. In this operation the lay shaft cover and reverse pinion shaft are also detached. Therefore, follow this up with removal of the first, second, third, fourth, and reverse gears.

Work on the Gear Shafts. It is best to entrust work on these components to a qualified agent, since the main cost of labour has been saved by undertaking the actual stripping oneself; while work on the synchromesh, etc., requires some special equipment which it is not economic to purchase. For the same reason, it is better to entrust actual reassembly of the box to a fully-equipped Reliant workshop.

CLUTCH, GEARBOX, AND TRANSMISSION

Propeller Shaft. Only the universals will require attention on the propeller shaft. Before the shaft is removed, jack up the vehicle and make pop marks on the mating flanges of the shaft and its couplings so that it can be refitted in its proper place. Then remove the nuts and bolts holding it to the couplings and lift away the shaft.

Clean the paint from the snap rings and bearing faces, so that the bearings can easily be extracted. Press the snap ring ends together and remove them by prising out with a screwdriver. Where a ring proves to be tight tap the bearing face lightly to relieve pressure on the ring.

With the rings out, support the shaft with one hand and, using a hide mallet, tap the radius of the yoke so that the bearing begins to emerge. Where it sticks, careful use of a short drift will enable you to tap the bearing from the inside. If you do this exercise great care, otherwise the race itself may be damaged.

Turn the yoke over and extract the bearings by hand. Don't lose any of the needles. Then repeat the operation on the other bearing, and the coupling flange can be removed from the spider.

Where wear has occurred a complete replacement unit should be fitted. If the original spider is reused, clean the parts thoroughly and assemble the needle rollers in the bearing races, using Vaseline or a very light grease to hold them. Place the spider in the yoke and tap the bearing into position, using a soft metal drift which is $\frac{1}{32}$ inch smaller than the diameter of the yoke hole. The bearing races must be a light driven fit in the yoke trunnions.

Before fitting the retainers coat the spider journal shoulders with shellac to make a good seal. If the joint binds slightly tap it gently with a hide mallet to relieve pressure.

Note that if the two halves of the propellor shaft are separated for any reason it is essential to reassemble them with the arrows on the universal joint and propellor shaft ends in line, so that uniform motion is transmitted. The shafts are separated by unscrewing the ring on the short end of the shaft. This section, when the shaft is offered up to the car, couples to the gearbox flange and the longer end to the axle flange coupling.

Rear Axle. Apart from the differential assembly and the crown wheel and pinion there is nothing in the rear axle to go wrong. Differential failure is unlikely, and since the average owner could not accurately mesh the crown wheel and pinion it is better not to attempt to strip this unit.

One job, however, which may be required is renewal of the pinion oil seal. To do this, jack up the front of the car and—for safety's sake—place blocks under the front wheel so that it is held securely.

Disconnect the propellor shaft from the axle coupling flange, and unscrew the four five-sixteenths of an inch bolts which hold the pinion housing. A hide or copper mallet can then be used to tap the pinion assembly out of the axle case.

Equip a vice with soft jaws and lock the pinion in it. Remove the split pin and undo the nut which holds the coupling flange to the pinion shaft, tapping it off with a mallet. Detach the pinion cap and tap out the oil seal, which is located below it. Carefully drift in the replacement, refit the cap, and replace the coupling flange. Tighten the nut, securing it with a new split pin, and replace the assembly on the axle. Finally, link up the propellor shaft again.

10 Steering and Suspension

ALL Reliants have a steering gear consisting of a single-stage worm, carried on the lower end of the column, in which engages a peg with a spherical end. Rotation of the column (and hence of the worm) causes the peg to turn and so move a rocker shaft. Two means of adjustment are provided—an adjuster screw set in the cover plate of the steering box, and shims placed between the lower end of the box and the end plate to control end play.

Stiff steering can be caused by the felt bush at the top of the column being too tight, by the steering tube being bent, or by partial seizure of the king pin owing to its greasing being neglected. Slack in the steering can almost invariably be traced to end play in the column.

End Float. Disconnect the track rod from the drop arm. Then, sitting in the driving seat, try to pull the wheel towards you. If you can feel any appreciable movement the end float is excessive, and the steering must be adjusted.

The end plate is bolted to the front of the steering box. Behind it there may be as many as half a dozen steel shims of identical thickness—0·04 to 0·05 inch. They are enclosed by a pair of paper washers, one on the box and one on the cover.

Remove one steel shim at a time, replacing and tightening the cover and checking the end float on each occasion before removing more. It should not be necessary to take off more than three shims, and it is essential not to remove so many that the steering becomes pre-loaded, for this could lead to damage to the adjustable ball race and to the ball tracks on the inner column. Note, also, that the correct thickness of the paper washers is 0·010 inch when new, falling to 0·005 to 0·007 inch when compressed.

Excessive backlash in the steering shows that the rocker shaft needs adjustment. This is carried out with the steering in the straight-ahead position. Release the adjuster lock nut slightly and gently screw down the adjuster until there is no play in the steering. Tighten the lock nut—still holding the adjuster as you do so—and do not release it until the lock nut is fully home.

Do not confuse the adjuster screw with the oil filler plug. The plug has a hexagon head, whereas the adjuster screw is slotted to take a screwdriver.

King Pin Renewal. If wear has taken place in the king pin to such an extent that the front wheel can be rocked about on its pivot, renewal is

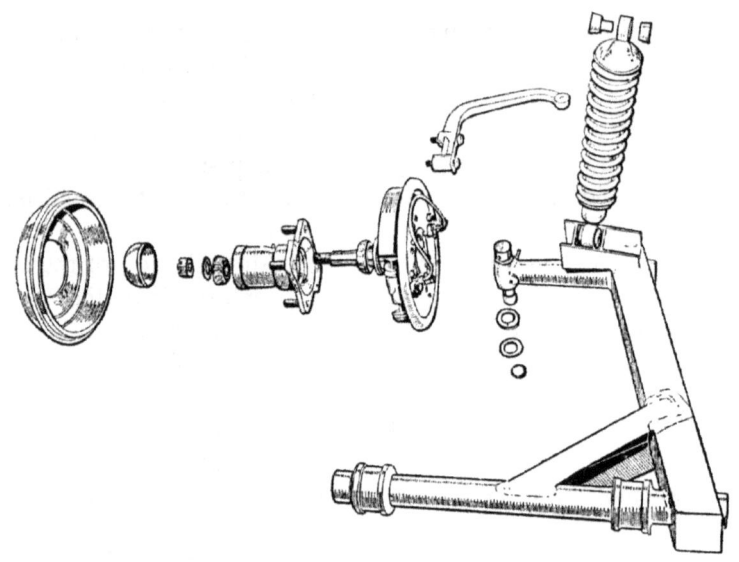

FIG. 43. FRONT SUSPENSION AND HUB
Strong but simple, this front suspension layout has been basically unchanged since the introduction of the Reliant, though the springing itself medium has altered.

vital. Jack up the car and block it up under the front engine mounting and radiator cross member. Then detach the front wheel, which is held by four $\frac{7}{16}$-inch nuts, and unscrew the $\frac{3}{16}$-inch round-headed metal thread screw which holds the hub cap. Tap the cap from the hub, unscrew the two countersunk screws which hold the brake drum to the hub, and slide the drum off.

From the stub axle take the large split pin, and undo the castellated nut. With a hide or copper mallet tap the hub from the shaft. The front tapered ball race will come away with it, and the rear one can be driven off independently. Take off the oil seal.

Unscrew the two self-locking nuts holding the brake assembly to the stub axle and the two bolts, nuts and the lock plate securing the steering arm to the brake plate and axle. Detach the brake shoes and—*without*

STEERING AND SUSPENSION 97

releasing the front hydraulic hose—slide the brake assembly onto the radius arm.

Two blind plugs are pressed into the stub axle. Remove them by drilling a small hole in each plug and easing it out with a drift. Then tap out the taper peg which locates the king pin and free the pin from the stub axle and radius arm. For this, you will need to use a steel drift slightly smaller in diameter than the king pin itself.

Slide the stub axle from the radius arm and, with a soft drift, tap out the old bushes. Two new Oilite bushes can then be driven in and reamed to size for a tight push fit ($\frac{5}{8}$-inch reamer). Then fit the stub axle to the radius arm with the new king pin. Drive home its taper peg, fit two new blind pegs, and rebuild the front brake assembly.

Refitting the Front Hub. Place the oil seal on its carrier and tap onto the stub axle the larger inner race from the two taper ball races. Next fit the outer race cages into the hub, the larger one going at the rear. Fill the hub with grease, slide it onto the stub axle, and follow it up with the smaller of the two inner races. A special drift is used to tap this into place.

Fit the D washer and the castellated nut. Tighten this as far as it will go, and then slacken it by one castellation only and fit a new split pin. Replace the front brake drum, the front hub grease nipple, and the hub cap, filled with grease. Secure it with its screw, and refit the front wheel.

Resetting Front Suspension. Front suspension resetting is a job which is applicable only to cars from Chassis No. 538001 to 562096. Jack up the front end and block it securely under the front cross tube. Then remove one of the $\frac{1}{4}$-inch bolts holding the rebound strap to the chassis.

Detach the two bolts which hold the torsion bar lever to the radius arm bracket and then, with a hide mallet, tap the lever from the bracket. It is held to the torsion bar by a taper pin. At the same time, remove the torsion bar from the splined abutment sleeve which is secured in the front cross member by four $\frac{5}{16}$-inch bolts. Leave these bolts in their holes and tap the lever until the torsion bar can be eased from the abutment sleeve by hand. Ensure that the splines are kept in line and you will avoid the need for several adjustments.

To raise the front of the car turn the torsion bar by one spline only. This will lift it by two inches, the adjustment being made in a clockwise direction. Turning the bar anti-clockwise will drop the front of the car.

The adjustment made, tap the torsion bar and lever into the radius arm bracket, secure the two bolts, and refit and secure the rebound strap. The car can now be lowered. On cars bearing Chassis Nos. 564001 to 580001 only a new front spring or complete damper unit can be used to compensate for sag at the front.

Rear Spring Removal. To remove one of the rear semi-elliptic springs jack up the car and block it under the chassis frame. Remove the appropriate rear wheel and release the shock absorber from the bottom spring bracket. From the U-bolts which hold the axle to the spring take the lock nuts and plain nuts, and then remove the short shackle bolt which secures the spring to the rear shackle.

Next undo the nut which holds the spring to the chassis and ease the spring off the pin. It is then free to be removed from the car. Immediately, mark the front end of the spring so that it can be properly reassembled. Reliants have their rear springs fitted off-centre, with the shorter end to the front.

Spring resetting is a job which must be entrusted to a forge which specializes in the heat treatment and tempering necessary for such work. Before fitting a newly-tempered spring, or a replacement spring, it is a good idea to grease it and wrap it in Drevo tape to seal it against road dirt and weather. Do not, however, cover the area where the axle and U-bolts must go.

To refit, merely reverse the sequence just described—but take care that when the shackle pins are replaced the lighting cable which runs along the inside of the chassis is not trapped between the shackle bolt and the frame.

Spring Bush Renewal. Detach the spring, as instructed above, and use a steel drift to drive out the old bush. Then press the new one into place. This can be done in a vice. If this facility is not available, obtain a short length of steel tube, a bolt (with some $3\frac{1}{2}$ inches of thread, long enough to pass through the bush, the spring, the eyelet, and the steel "collar"), and washers to fit the bolt. Slip one washer over the bolt and add the steel collar. Butt its edges against one side of the spring eyelet and feed the bolt through the old bush. At the other side, butt the new bush against the old and pass the bolt through its centre. Add the washer and the nut. If the nut is now tightened the old bush will be pressed out and the new one pulled in at the same time.

Rear Dampers. No adjustment is possible to the rear dampers. Where the action of a damper is unsatisfactory it must be replaced with a new unit. The damper is held at the chassis end by a three-eighths of an inch self-locking nut with a plain washer, and its bottom end is held to the rear spring anchor plate by a $\frac{3}{8}$-inch bolt with a plain washer and a self-locking nut.

11 Looking after the Brakes

THE Reliant has a conventional braking system, comprising three drum-type brakes with hydraulic operation from a single pedal. In addition, there is a mechanical handbrake linkage which operates on the rear wheels only.

Very little maintenance or overhaul work is required here. The hydraulic system must, however, be checked regularly for signs of leakage or deterioration of the hoses. From time to time the system must also be "bled" to exclude unwanted air from the fluid, and every two years it is advisable to flush the system thoroughly. The position of the shoes relative to the brake drums must be adjusted as wear takes place, and the handbrake linkage must be set and lubricated. Once the shoe linings have become really worn the shoes must be removed and replaced by freshly-lined exchange parts. No other work should be necessary, unless there is a failure of a hose or of the seals in any of the operating cylinders.

Brake Bleeding. If in use the brake pedal develops a soft "spongy" feel, and it becomes necessary to "pump" the pedal to obtain adequate pressure on the shoes, it is a sure indication that the brake lines require bleeding to rid the fluid of air which has managed to enter.

You will need for the job a clean jar—really clean, for no dirt can safely be allowed near the sensitive braking system—and a tin of Girling brake fluid, together with a short length of rubber piping.

Fit the piping over the bleed nipple on the rear nearside brake back-plate, and pour into the jar sufficient fluid to enable the end of the pipe to be immersed to a depth of about half an inch or so.

Now get an assistant to sit in the car to operate the brake pedal. Chock the wheels and release the handbrake. Then, while the assistant depresses the brake pedal with a long steady stroke open the bleed screw about three-quarters of a turn. Fluid will shoot out into the jar and, initially, you will note that it contains air bubbles. The pedal should continue to be pressed and released slowly, each stroke taking place at about three-second intervals, until the fluid expelled from the pipe no longer contains air. Then, during the next down stroke of the pedal tighten the bleed screw. Then move to the front of the car and repeat the process with the front bleed nipple.

It is essential, during bleeding, to ensure that the level of fluid in the hydraulic reservoir does not fall so low that air can be drawn into the pipe. Therefore, check that the reservoir is full before starting, and top it up again before moving on to the front brake. Make a final check after bleeding is completed, and if the fluid level is below the normal

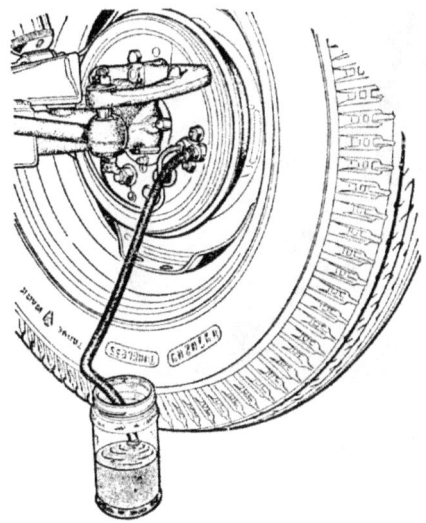

FIG. 44. BRAKE BLEEDING

The front brake ready to be "bled," with the bleed nipple connected to a jar of fluid by means of a rubber pipe. The end of the pipe must be completely submerged during the whole operation.

mark top it up once more. Note that only Girling crimson brake fluid must be used, otherwise the seals will be damaged.

Providing the jar in which the fluid was caught is absolutely clean the old fluid may be used again once it has stood for a few days to allow it to de-aerate itself. But many owners prefer not to take a chance on it for the sake of the few shillings involved, and merely pour all old fluid away as a matter of course.

Flushing the System. For a complete flush out the procedure is basically similar to that used for bleeding. In this instance, however, the process is continued until all the fluid has been expelled from both bleed nipples —and, of course, there is no need to start off with fluid in the jar.

When the system is empty it is filled with denatured alcohol (methylated spirits) which is pumped through in a manner similar to that used for draining the original fluid. When all the meths has been expelled a second flushing must be carried out, using fresh Girling brake fluid. This is vital, since some of the meths will remain in the master cylinder and this *must* be scoured out, otherwise the brakes will fail to operate properly.

Keep operating the pedal until all the fluid put in for the second flushing

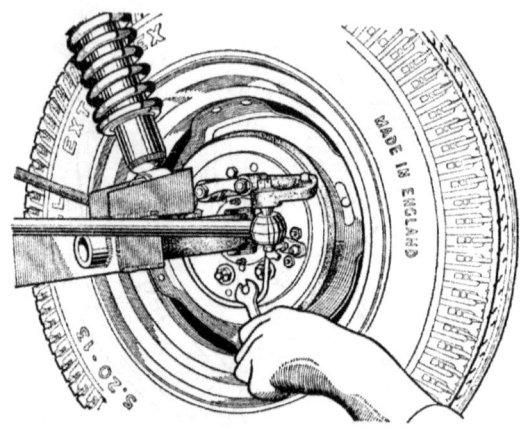

FIG. 45. FRONT BRAKE ADJUSTMENT
Each front brake shoe is independently positioned by turning its hexagon-headed adjuster.

has been pumped out again. Then refill the system with brake fluid, and bleed the brakes to complete the job.

Front Brake Adjustment. Apply the handbrake and jack up the car so that the front wheel clears the ground. Fully release both the hexagon-headed adjuster bolts on the back plate. This is done by turning them anti-clockwise. Select an adjuster, and turn it clockwise until the brake shoe which it controls is touching the drum. You can feel this by gently rocking the wheel. Now back off the adjuster again until the wheel just turns freely. Repeat this procedure with the second adjuster, and finally spin the wheel to ensure that the brakes do not touch at any point. Then lower the car.

Rear Brake Adjustment. The principle is the same as for the front brake, but a single adjuster serves both shoes in each rear brake. Release the handbrake, chock the opposite wheel and the front wheel, and raise one

LOOKING AFTER THE BRAKES

side of the car. Then turn the squared end of the adjuster clockwise until resistance is felt. Now slacken it off by two "clicks," whereupon the wheel should rotate freely. The only exception to this is when new shoes have been fitted, then the adjuster is released by three "clicks" to allow for possible lining expansion, the closer adjustment being adopted once the linings have settled down.

Lower the car, transpose the chocks, raise the other wheel, and make the same adjustment on the other side.

Handbrake Adjustment. Normally, the adjustment outlined above will serve to keep the handbrake in operating trim also, and the length of the

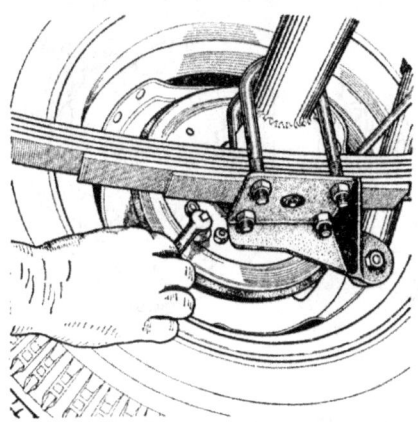

FIG. 46. REAR BRAKE ADJUSTMENT
One adjuster on the backplate controls both shoes in a rear brake.

cable connecting the handbrake linkage to the lever should not therefore be varied.

If the cable has been renewed, however, it may be that some adjustment may need to be made. It is done by locking the rear shoes to the drums by turning the adjuster with the handbrake in the "off" position. The cable can then be adjusted by means of the clevis fitted on the adjuster screw at the rear of the cable. This should be turned until the clevis pin can be slipped easily through the clevis and balance lever.

Replacement of the brake cable is effected by removing the split pin and clevis pin holding the cable to the cable lever, and then the split and clevis pins holding the cable to the balance lever. That done the clevis is unscrewed from the cable and the handbrake is removed, and the new cable fitted by reversing the sequence.

Fitting Brake Shoes. When the shoes become worn ready-lined replacements should be fitted. Besides being equipped with the correct linings these are also accurately ground to size, and this ensures that the bedding-in process is rapidly completed.

To reline the front brake, jack up the car and remove the wheel and the brake drum, which is held to the hub by a pair of countersunk screws. This bares the brake backplate, on which the shoes are carried. With a large screwdriver prise one shoe out of the groove in the wheel cylinder

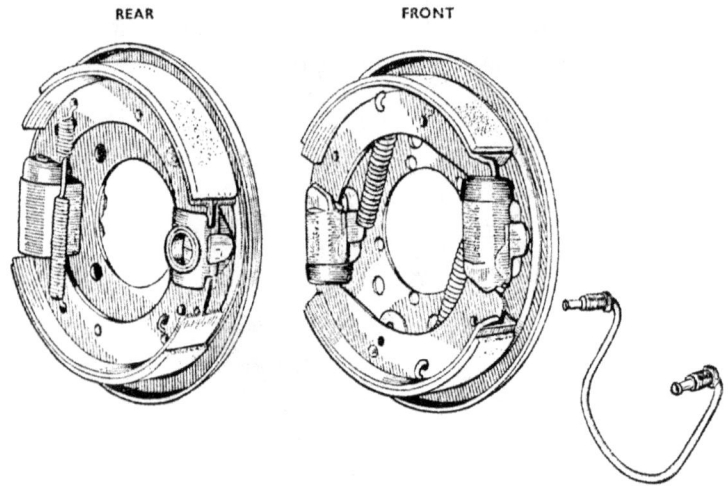

FIG. 47. THE BRAKE BACKPLATES

The rear brake, of leading-and-trailing shoe type, has a single slave cylinder. The two-leading shoe front brake has a pair of cylinders. Note the positioning of the shoe pull-off springs.

piston in which its end rests, and then pull both shoes away from the plate. Fix a strong elastic band over the wheel cylinder so that there is no chance of a piston coming out. Note that the brake pedal should never be operated with the shoes removed—the pressure applied would be sufficient to shoot the pistons from their cylinders.

Clean the backplate and check the exterior of the slave cylinders for signs of leakage. Take the chance of testing the shoe adjusters for easy working, and slacken them anti-clockwise to their fully-off position. Where lubrication is required give it, using Girling brake grease.

Replacement shoes should always have new pull-off springs. Attach these to the shoes and offer them up to the back plate. The half-round slot on the end of each shoe butts against the pivot pin; the other is in contact with the wheel cylinder. Position one shoe first, and then use a screwdriver under the web of the remaining shoe to lever it into contact with its cylinder. Ensure that the brake drums are free from grease and dust, and refit them. Replace the wheel, and adjust the brake as described earlier in this chapter.

The replacement of shoes in the rear brakes is a similar operation, though with one or two variations. The rear brake shoes have a common slave cylinder, and on removing the shoes one must first be lifted out of engagement with the adjuster link and the cylinder piston. On replacement the shoes are linked by springs the shorter of which is fitted at the adjuster end and on the side of the webs facing the back plate. One shoe is first located in the adjuster link and slave cylinder slots, and the other is then prised into place. That done, the rubber band is removed from the cylinder. After replacement of the drum and the wheel, adjustment is carried out in the normal way.

12 Chassis, Body, and Tyres

ALL Reliant threewheelers are built around a sturdy box-section chassis, to which the body is bolted. And most of the Reliants on the road today have durable glass-fibre bodies which require little maintenance other than routine cleaning and polishing.

On the chassis itself there is virtually no routine work to be done other than that already detailed under the various separate headings. However, it may in some instances be helpful to understand how some of the ancillaries are attached.

Fuel Tank. The tank is fitted with its units already in place. It is held on brackets welded to the main side members, the half-inch tank rod being slipped through the brackets and the centre of the tank and held with nuts and split washers. Then two half-inch by quarter-inch UNF bolts, with spring washers, are screwed through the brackets and into bosses on the tank and tightened down. Finally, the two nuts on the centre rod are fully secured. That done, the petrol pipe is fitted and held to the offside chassis member with a quarter-inch spring clip. It is also clipped to the gearbox.

Brake Master Cylinder. This is bolted to a bracket welded to the chassis. It is held in place by a pair of $1 \times \frac{5}{16}$-inch bolts, each with a plain washer, a spring washer, and a $\frac{5}{16}$-inch nut.

Exhaust System. The exhaust system is in three sections, fitted as two units. The down-pipe has the silencer ready-attached and it is bolted to the manifold by four $1 \times \frac{1}{4}$-inch bolts with spring washers and brass quarter-inch nuts. A copper gasket is interposed at the joint. The silencer is supported by a short flexible strap joining a bracket formed on the silencer to one welded to the front cross-member. This strap is held at each end by a $\frac{5}{8} \times \frac{1}{4}$-inch bolt with self-locking nut.

From the silencer a tail pipe carries the exhaust gases to the rear of the car. At the front the pipe is clipped into the silencer, and at the rear it is supported by a flexible strap bolted to a tail-pipe bracket with a $\frac{5}{8} \times \frac{1}{4}$-inch bolt and self-locking nut. Its other end is fixed to a tapped plate welded to the main chassis member, the fixing being by means of a 1 × ¼-inch steel plate through which passes a ⅝ × ¼-inch bolt with spring washer.

Radiator. The radiator is held on studs by $\frac{5}{16}$-inch self-locking nuts. Its mounting is semi-flexible, with cushioning by means of thick rubber washers backed by $\frac{5}{16}$-inch coach washers.

Stabilizer Rod. A rubber grommet is inserted into the bracket secured to the rear engine mounting plate. Two $\frac{5}{16}$-inch nuts are used on the rod,

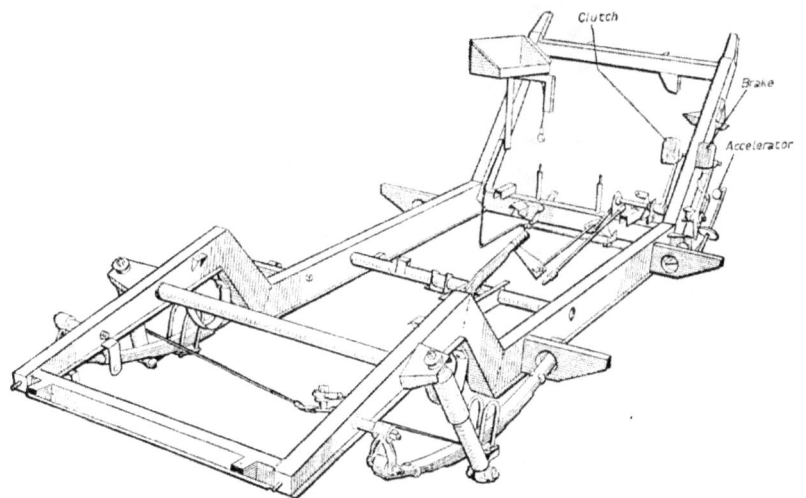

FIG. 48. THE RELIANT CHASSIS

Backbone of all Reliants is a sturdy chassis like this. Note the disposition of the rear suspension units and the mounting of the controls on a cross shaft.

together with a special washer, and the rod is slipped through the grommet. The nuts are then loosely attached. The eye end of the rod is fitted to the bottom stud of the offside engine mounting rubber and held with ⅜-inch self-locking nuts. Tensioning is done with the $\frac{5}{16}$-inch nuts on each side of the engine plate, the rod being tightened until all slackness is removed. The additional $\frac{5}{16}$-inch nuts are then tightened down to lock the rod in place.

Windows, Removal and Fitting. Reliant front door windows are in two halves, the forward part being fixed while the rear half slides in a felt-lined channel. They have a common rubber surround, which is itself locked into position by an inner rubber sealing strip.

108   THE BOOK OF THE RELIANT

To detach the windows this sealing strip must be removed by inserting a thin knife or screwdriver into the joint and easing it out till enough

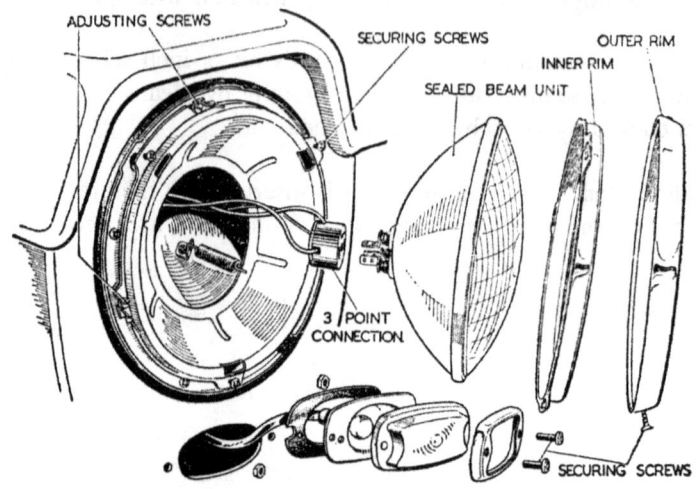

Fig. 49. Head-Lamp and Side-Lamp Fixing
How the head and side lamps of the Regal 3/25 are fixed into the body.

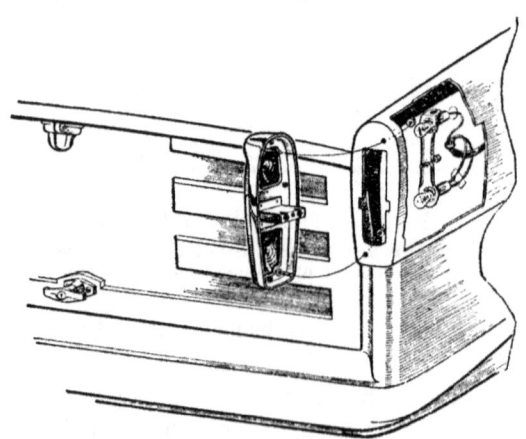

Fig. 50. Rear-Lamp Cluster
The disposition of the lamps and the method of fixing the lens units at the rear of the Regal 3/25.

is free to enable it to be gripped by hand. A gentle pull will then suffice to pull the entire locking strip away.

Detach the thin rubber strip which separates the window from the casing and press gently on the glass, from the inside, to remove the window. Ease the glass from the weatherstrip with the aid of a piece of tapered wood. Slide the window locking catch from beneath the sliding window's channel and that glass can be taken out.

Windows in hard top models are dealt with in the same way. To refit. insert the glass into the surround doing the job by hand as far as possible,

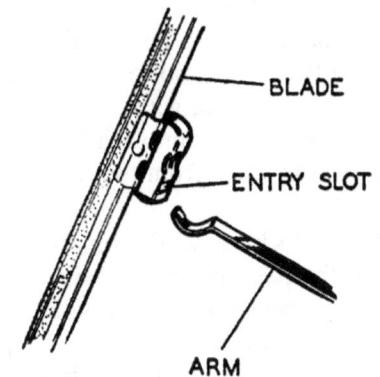

FIG. 51. WIPER BLADE ATTACHMENT

Each wiper blade is held on its arm by means of a hook and slot. To free the blade pull the arm forwards, and merely unhook the blade unit.

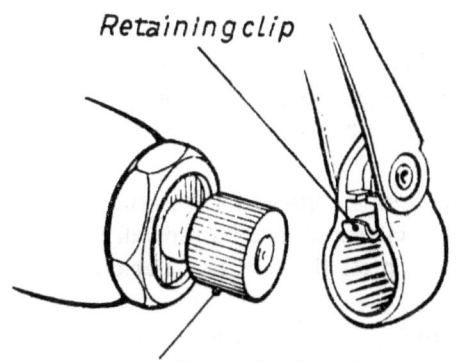

FIG. 52. WIPER ARM ATTACHMENT

If the entire wiper arm has to come off, tilt the arm forwards and lift the retaining clip clear of the driving drum. The arm will then slide off its splines.

Complete it by using a thin piece of tapered softwood to ease the rest of the surround over the glass. Then refit the locking rubber. Most of it will go in by hand, and the rest can be inserted by careful use of a screwdriver.

Precisely the same method is used for windscreen replacement. If the original screen has been shattered all loose glass must be removed from the surround, and the wiper blades will in any case have to be taken off their spindles.

Leaks in the surrounds can be cured by injecting permanent sealing compound between the surround and the glass. Any which is squeezed onto the glass itself can be removed by use of a rag soaked in petrol.

Door Removal. It is inadvisable to remove the door hinges once these have been positioned by the factory, so to detach a door first remove the check strap at the bottom and then, with a steel drift 0·010 inch smaller in diameter than the hinge pin, tap out each hinge pin until it clears the hinge. The door can then be lifted from the car.

Door Locks. Both the locking handle on the driver's door and the non-locking handle on the passenger's door are held by two chromium-plated wood screws. The locks are secured by four black round-headed wood screws and two chromium screws.

Six three-quarter-inch No. 6 wood screws hold the male dovetail to the door, while the female dovetail is held to the door slam pillar by four $\frac{3}{4}$-inch No. 6 screws. The slam locking plate is held by two countersunk wood screws.

Bonnet and Boot Lids. To detach the bonnet lid simply remove the nuts which hold it to its hinges. The same applies to the boot lid, whose lock is, in addition, held by a pair of wood screws.

Radiator Grille. Spring clips secure the grille to the body. To remove it, merely locate the clips and insert a screwdriver at each point to lever them clear. To refit, press the clips home again.

Foot Wells. Both foot wells are held by means of $\frac{3}{16}$-inch metal thread screws and nuts at the fascia panel, by wood screws to the hinge pillar, and by recess-headed screws and spire nuts to the footboards.

Removal entails, first, releasing the screws from the hinge pillar, next the floor fittings, and lastly the fascia screws. That done, the fascia's own screws are removed so that it can be raised and so allow the footwell to slide clear.

Fascia Panel. Eight chromium-plated countersunk wood screws hold the fascia panel to the bottom windscreen door rail and the door pillars. To detach it remove the plated locking strip from the instrument panel,

CHASSIS, BODY, AND TYRES

lift the bonnet lid, remove the instrument panel nuts, and then slide the instrument panel through the aperture in the fascia.

Next, detach the wood screws from the bottom rail of the screen and from the door pillars, remove the foot well and engine cowl bolts and lift the fascia panel clear. Refitting involves only a simple reversal of this procedure.

Seats, Adjustment and Removal. To adjust the front seats slacken the thumb screw which locates the bottom framework of the seat and slide it backwards or forwards to obtain the most comfortable position. Then relock the screw.

Front seat removal is effected by releasing the bolts holding its three brackets. The nuts are beneath the floorboards. Detach them, and drive the bolts upwards. Remove the brackets (and their locking plates from below the car) and take out the seats.

The rear seat back rest and cushion are held in place by press studs.

Engine and Prop-shaft Covers. Eight Phillips screws and spire nuts hold the propellor-shaft cover to the floorboards.

The three-piece engine cover has its top half secured to the facia panel by a pair of $\frac{3}{16}$-inch metal thread screws, nuts, and spring washers, and two more of these hold it to the prop-shaft cover. The two engine cover side panels are held by spring clips.

Puncture Repair. The Reliant is equipped with tubeless tyres, which have a high degree of resistance to puncturing. Normally, a self-sealing effect is obtained when such a tyre is holed, and provided the cause of the puncture remains in the cover it is possible to carry on for some while before repairs are made.

A special tubeless tyre repair kit has to be used. This contains a plugging needle, various rubber plugs, and rubber solution. Take off the wheel and remove the nail or other foreign body from the tyre and dip the plugging needle into the solution. Then insert it into the cut and move it up and down to coat the hole with solution, repeating this until the hole is well covered. Then take a plug about double the size of the hole and stretch and roll it into the plugging needle eye. Dip the plug into the solution and insert the needle and plug into the hole. Thrust it in so that a good section of the plug is left inside the tyre and then withdraw the needle. Cut off the projecting section of the plug about an eighth of an inch from the tread surface, and reinflate the tyre before replacing the wheel.

Appendix Facts and Figures

1. SIDE-VALVE ENGINES

Capacity	C.R.	Firing order	Tappet clearance	Ring gap	Valve seat angle/width	Valve spring free length
747·5 c.c.	5·7:1	1–3–4–2	0·004 inch cold	0·005 inch	45°/$\frac{1}{32}$-in.	1·52 inch ± 0·031 inch

2. OVERHEAD-VALVE ENGINES

Capacity	C.R.	Firing order	Tappet clearance	Ring gap	Valve seat angle/width	Valve spring free length
598 c.c.	7·8:1	1–3–4–2	0·006 inch cold	0·006—0·008 inch	45°/$\frac{1}{32}$-in.	1·50 inch ± 0·031 inch
700 c.c.	8·4:1	,,	,,	,,	,,	,,

3. TORQUE WRENCH SETTINGS, SIDE-VALVE ENGINES

Cylinder head nuts	Big end bolts	Centre main bearing bolts	Front and rear main bearings	Flywheel bolts
230/260 lb-in.	250 lb-in.	300/350 lb-in.	310 lb-in.	250 lb-in.

4. TORQUE WRENCH SETTINGS, OVERHEAD-VALVE ENGINES

Cylinder head nuts	Big end bolts	Centre main bearing bolts	Front and rear main bearings	Flywheel bolts
260/280 lb-in.	250 lb-in.	350 lb-in.	350 lb-ins.	350 lb-in.

FACTS AND FIGURES

5. CAPACITIES, SIDE-VALVE MODELS

Fuel	Sump	Cooling	Gearbox	Rear Axle
6 gal	5 pints	9 pints	1 pint	2 pints

6. CAPACITIES, OVERHEAD-VALVE MODELS

Fuel	Sump	Cooling	Gearbox	Rear Axle
6 gal	7 pints	6 pints*	1 pint	2 pints

* 7 pints when a heater is fitted.

7. RECOMMENDED LUBRICANTS, ALL MODELS

Lubricant Brand	B.P.	Esso	Mobil	Shell	Wakefield
Engine (S.V.) summer and winter	Energol SAE 30	Esso Extra 20W/30	Mobiloil "A"	Shell X-100 30	Castrol XL
Engine (O.H.V.) summer and winter	Energol 20 or Viscostatic	Esso Extra 20W/30	Mobiloil Arctic or Special	Shell X-100 20 or Multi-grade 10W/30	Castrolite
Gearbox (S.V.) summer and winter	Energol SAE 90	Esso Gear Oil 90	Mobilube C 90	Spirax 90	Castrol D Gear oil 90
Gearbox (O.H.V.) summer and winter	All brands as for use in the engine (o.h.v. model)				
Rear axle (all models) summer winter	Energol SAE 90 EP	Esso Gear Oil 90	Mobilube GX 90	Spirax 90 EP	Castrol Hypoy
Steering box	Energol SAE 90 EP	Esso Gear GP90	Mobilube GX90	Spirax 90 EP	Castrol Hypoy
Front hub (all models)	Energrease L 21M	Multi-purpose Grease H	Mobilgrease MP	Retinax A	Castrolease LM

8. PERIODIC LUBRICATION

Period	Application	Operation	Lubrication Point	
			Number	Position
Every 1,000 miles	Steering	Grease Gun	2 each 2 2	Track Rod. Linkage Swivel Pin. Front Hub Swivel Pin.
	Brake Cross Shaft	Grease Gun	1	Outer end of Brake Cross Shaft.
	Hand Brake	Grease Gun	2 (o.h.v., 1)	Brake Rod Compensator.
	Propellor Shaft	Grease Gun	1 2	Propellor Shaft Splines. Universal Joints.
	Brake and Clutch Linkage	Oil Can		Clevis Pins, Return Springs, all moving joints.
	Throttle Linkage	Oil Can	3	Throttle Cross Shaft Bearings. All moving joints.
	Steering Box	Oil	Filler Plug	Top up if necessary through filler plug on inside face of steering box.
	Rear Road Springs	Oil Spray or Brush		Remove mud and dirt and brush or spray with penetrating oil.
Every 3,000 miles	Front Wheel Bearings	Grease Gun	1	Hub. Do not overcharge or grease may penetrate through to brake linings.
	Air Cleaner	Oil		Remove, wash in petrol — dry — dip in oil.
As Required	Brake Fluid Reservoir			Top up as necessary.
	Body Parts	Oil Can		Door and Bonnet and Boot Hinges.

9. READY REFERENCE

Tyre sizes: 5·20 × 13 in.

Tyre pressures: All s.v. models, 20/22 psi front, 22/24 psi rear. All o.h.v. models, 18 psi front, 20/24 psi rear. Vans, 20 psi front, 24/30 psi rear.

Contact-breaker gap: 0·015 in.

Ignition timing: 10° B.T.D.C. (fully retarded)

Recommended plugs and gaps: All s.v. models, Autolite AE6 (0·020/0·025 in.). All o.h.v. models, Autolite AG42 (0·025 in.).

Fuel tank capacity: 6 gal

Unladen dry weight: 7 cwt 100 lb

NOTES

Index

ADVANCE/RETARD mechanism, 78
Air cleaner, 70

BACKLASH, 95
Battery—
  care, 88
  level, 86, 87, 88
  specific gravity, 87, 88
Big-end bearings—
  o.h.v. models, 62
  s.v. models, 54, 57
Brakes, 41
  adjusting, 102, 103
  bleeding, 100, 101
  flushing, 101
  shoes, 104, 105

CAMSHAFT, s.v. models, 55, 57, 58
Carburettor, overhaul, 67, 68
Chassis, 107
Clutch, 39, 40
  removal, s.v. models, 53, 89
  renovation, 89, 90
  3/25, 62
Contact-breaker, 37, 38, 76, 77
  gap, 115
Cooling, 64
Crankshaft—
  o.h.v. models, 63
  s.v. models, 55, 57, 58
Cylinder block, s.v. models, 52, 57
Cylinder head—
  o.h.v. models, 59, 60
  s.v. models, 44, 50

DAMPERS, 99
Decarbonizing, 42, 43, 58
Distributor, 38, 39, 51, 76, 77
  overhaul, 77, 78
Doors, 110
Dry weight, 115
Dynamo—
  armature, 81
  brushgear, 80, 82, 83
  end bracket, 83

  lubrication, 79
  testing, 82

ECONOKIT, 67, 68
Engine—
  assembly, 3/25, 63
  decarbonizing 3/25, 58
  removal, 53, 61
  s.v. models, 53
  stripping 3/25, 61, 62, 63
  s.v. models, 54
Exhaust pipe, 106

FAN belt, 40, 41
Fascia, 110, 111
Fault tracing, 18, 19, 20, 21, 23, 27
Front hub, 97
Fuel—
  capacity, 115
  pump, 43, 44, 51, 70, 72, 73, 74, 75
  system, 66

GEARBOX—
  dismantling, 90, 91, 92
  removal, s.v., 53, 89
  3/25, 61

IDLING, adjusting, 68, 70
Ignition timing, 115

JETS, 68, 70

KING pin, 96, 97

LAMPS, 108
Lids, 110
Liners, o.h.v. models, 63
Locks, 110
Lubrication, 30, 31, 32, 33, 34, 35, 36, 113, 114

MAIN bearings—
  o.h.v. models, 63
  s.v. models, 56

117

Maintenance—
　daily, 29
　weekly, 29, 30
Manifold, s.v. models, 43
Master cylinder, 106

OIL glaze, 52
Oil pump, s.v. models, 54
Oils, recommended, *see* Lubrication

PISTONS, 44, 45, 52
Plugs, 115
Propellor shaft, 93
Punctures, 111
Pushrods, 58

RADIATOR—
　cap, 64
　fixing, 107
Rear axle, 35, 93, 94
Rear spring—
　bush renewal, 98
　removal, 98
Reverse flushing, 64
Rockers, 58

SEATS, 111
Silencer, 106
Sparking plugs, gap, 115
　recommended, 115

Stabilizer rod, 107
Starter motor—
　brushes, 85, 86
　rebuilding, 86
　removal, 83, 84
　stripping, 84, 85
Steering—
　box, 95
　column, 95
　end float, 95
Suspension, resetting, 97

TAPPETS, 36, 37, 51
Timing chain, s.v. models, 57
Tools, 15, 16, 17
Torsion bars, 97
Tyre pressures, 115
　sizes, 115

VALVE springs, 48, 49
Valves, 45, 46, 47, 49
　grinding, 59
　replacing, 60

WINDOWS, 107, 108
Windscreen wipers, 109
Wiring, 22, 24, 25, 26

VELOCEPRESS MANUALS - MOTORCYCLE BY MAKE

AJS 1932-1948 SINGLES & TWINS 250cc THRU 1000cc (BOOK OF)
AJS 1945-1960 SINGLES 350cc & 500cc MODELS 16 & 18 (BOOK OF)
AJS 1955-1965 SINGLES 350cc & 500cc (BOOK OF)
AJS 1957-1966 FACTORY WSM - ALL SINGLES & TWINS
ARIEL UP TO 1932 (BOOK OF)
ARIEL 1932-1939 PREWAR MODELS (BOOK OF)
ARIEL 1933-1951 (WORKSHOP MANUAL)
ARIEL 1939-1960 4 STROKE SINGLES (BOOK OF)
ARIEL 1958-1964 LEADER & ARROW (BOOK OF)
BMW R26 R27 (1956-1967) FACTORY WORKSHOP MANUAL
BMW R50 R50S R60 R69S (1955-1969) FACTORY WORKSHOP MANUAL
BRIDGESTONE 90 SERIES FACTORY WSM & PARTS CATALOGUE
BRIDGESTONE 175 SERIES FACTORY WSM & PARTS CATALOGUE
BRIDGESTONE 350 SERIES FACTORY WSM & PARTS CATALOGUES
BSA SERVICE SHEETS MASTER CATALOGUE ALL MODELS 1945-1967
BSA BANTAM D1 TO D7 1948-1966 FACTORY SERVICE SHEETS MANUAL
BSA BANTAM ALL MODELS FROM 1948 ONWARDS (BOOK OF)
BSA BANTAM D14 FACTORY WORKSHOP & INSTRUCTION MANUAL
BSA SINGLES & V-TWINS UP TO 1927 (BOOK OF)
BSA SINGLES & V-TWINS UP TO 1930 (BOOK OF)
BSA SINGLES & V-TWINS UP TO 1935 (BOOK OF)
BSA SINGLES & V-TWINS 1936-1939 (BOOK OF)
BSA C10, C11 & C12 1945-1958 FACTORY SERVICE SHEETS MANUAL
BSA OHV & SV SINGLES 250-600cc 1945-1959 (BOOK OF)
BSA C15 & B40 1958-1967 FACTORY SERVICE SHEETS MANUAL
BSA OHV & SV SINGLES 250cc (ONLY) 1954-1970 (BOOK OF)
BSA B31, B32, B33 & B34 1945-60 FACTORY SERVICE SHEETS MANUAL
BSA OHV SINGLES 350 & 500cc 1955-1967 (BOOK OF)
BSA M20, M21 & M33 1945-1963 FACTORY SERVICE SHEETS MANUAL
BSA TWINS A7 & A10 1948-1962 FACTORY SERVICE SHEETS MANUAL
BSA TWINS A7 & A10 1948-1962 (BOOK OF)
BSA TWINS A50 & A65 1962-1965 FACTORY WORKSHOP MANUAL
BSA TWINS A50 & A65 1962-1969 (SECOND BOOK OF)
DOUGLAS 1929-1939 PREWAR ALL MODELS (BOOK OF)
DOUGLAS 1948-1957 POSTWAR ALL MODELS FACTORY SHOP MANUAL
DUCATI 160cc, 250cc & 350cc OHC MODELS FACTORY SHOP MANUAL
HONDA 50 ALL MODELS UP TO 1970 INC MONKEY & TRAIL (BOOK OF)
HONDA 90 ALL MODELS UP TO 1966 (BOOK OF)
HONDA 125-150cc TWINS C/CS/CB/CA FACTORY WORKSHOP MANUAL
HONDA 250-305 TWINS C/CS/CB FACTORY WORKSHOP MANUAL
HONDA 450 CB/CL 1965-1974 K0 TO K7 WORKSHOP MANUAL
HONDA C100 SUPER CUB FACTORY WORKSHOP MANUAL
HONDA C110 SPORT CUB 1962-1969 FACTORY WORKSHOP MANUAL
HONDA TWINS & SINGLES 50cc THRU 305cc 1960-1966 (BOOK OF)
HONDA TWINS ALL MODELS 125cc THRU 450cc UP TO 1968 (BOOK OF)
INDIAN PONYBIKE, BOY RACER & PAPOOSE ILL PARTS LIST & SALES LIT
J.A.P. ENGINES 1927-1952 & MOTORCYCLES 1934-1952 (BOOK OF)
MATCHLESS 1931-1939 ALL MODELS 250cc THRU 990cc (BOOK OF)
MATCHLESS 1945-1956 350 & 500cc SINGLES (BOOK OF)
MATCHLESS 1955-1966 350 & 500cc SINGLES (BOOK OF)
MATCHLESS 1957-1966 FACTORY WSM - ALL SINGLES & TWINS
NEW IMPERIAL ALL SV & OHV FROM 1935 ONWARDS (BOOK OF)
NORTON 1932-1939 PREWAR MODELS (BOOK OF)
NORTON 1932-1947 (BOOK OF)
NORTON 1938-1956 (BOOK OF)
NORTON 1955-1963 MODELS 19, 50 & ES2 (BOOK OF)
NORTON 1955-1965 DOMINATOR TWINS (BOOK OF)
NORTON 1960-1970 TWIN CYLINDER FACTORY WORKSHOP MANUAL
NORTON 1970-1975 COMMANDO FACTORY WORKSHOP MANUAL
NORTON 1975-1978 MK 3 COMMANDO FACTORY WORKSHOP MANUAL
PANTHER 1932-1958 LIGHTWEIGHT MODELS 250 & 350cc (BOOK OF)
PANTHER 1938-1966 HEAVYWEIGHT MODELS 600 & 650cc (BOOK OF)
RALEIGH MOTORCYCLES 1919-1933 (BOOK OF)
ROYAL ENFIELD 1934-1946 SINGLES & V TWINS (BOOK OF)
ROYAL ENFIELD 1937-1953 SINGLES & V TWINS (BOOK OF)
ROYAL ENFIELD 1946-1962 SINGLES (BOOK OF)
ROYAL ENFIELD 1958-1966 250cc & 350cc SINGLES (SECOND BOOK OF)
ROYAL ENFIELD 736cc INTERCEPTOR FACTORY WORKSHOP MANUAL
RUDGE 1933-1939 (BOOK OF)
SUNBEAM 1928-1939 (BOOK OF)
SUNBEAM 1946-1957 S7 & S8 (BOOK OF)
SUZUKI 50cc & 80cc UP TO 1966 (BOOK OF)
SUZUKI T10 1963-1967 FACTORY WORKSHOP MANUAL
SUZUKI T20 & T200 1965-1969 FACTORY WORKSHOP MANUAL
SUZUKI TWINS 1962 ONWARDS 125-500cc WORKSHOP MANUAL
TRIUMPH 1935-1939 PREWAR MODELS (BOOK OF)
TRIUMPH 1935-1949 (BOOK OF)
TRIUMPH 1937-1951 (WORKSHOP MANUAL)
TRIUMPH 1945-1955 FACTORY WORKSHOP MANUAL
TRIUMPH 1945-1958 TWINS (BOOK OF)
TRIUMPH 1956-1969 TWINS (BOOK OF)
VELOCETTE 1925-1970 ALL SINGLES & TWINS (BOOK OF)
VILLIERS ENGINE UP TO 1959 INC. 3 WHEELERS (BOOK OF)
VILLIERS ENGINE UP TO 1969 (BOOK OF)
VINCENT 1935-1955 (WORKSHOP MANUAL)
YAMAHA 1961-1967 YA5 & YA6 (WORKSHOP MANUAL & ILL PARTS LIST)
YAMAHA 1971-1972 JT1& JT2 (WORKSHOP MANUAL & ILL PARTS LIST)

VELOCEPRESS TECHNICAL BOOKS - MOTORCYCLE

1930'S BRITISH MOTORCYCLE CARBS & ELEC COMPONENTS (BOOK OF)
1930'S BRITISH MOTORCYCLE ENGINES (OVERHAUL & MAINTENANCE)
1930'S BRITISH MOTORCYCLE GEARBOXES & CLUTCHES (BOOK OF)
CATALOG OF BRITISH MOTORCYCLES (1951 MODELS)
LUCAS ELECTRONICS BRITISH M/CYCLES REPAIR & PARTS (1950-1977)
MOTORCYCLE ENGINEERING (P.E. Irving)
MOTORCYCLE ROAD TESTS 1949-1953 (Motor Cycle Magazine UK)
SPEED AND HOW TO OBTAIN IT (Motor Cycle Magazine UK)
TUNING FOR SPEED (P.E. Irving)

VELOCEPRESS MANUALS - SCOOTERS BY MAKE

BSA SUNBEAM SCOOTER WORKSHOP MANUAL 1959-1965
BSA SUNBEAM SCOOTER 1959-1965 (BOOK OF)
LAMBRETTA 1947-1957 ALL 125 & 150cc MODELS (BOOK OF)
LAMBRETTA 1957-1970 LI & TV MODELS (SECOND BOOK OF)
NSU PRIMA 1956-1964 ALL MODELS (BOOK OF)
TRIUMPH TIGRESS SCOOTER WORKSHOP MANUAL 1959-1965
TRIUMPH TIGRESS SCOOTER (BOOK OF)
VESPA 1951-1961 (BOOK OF)
VESPA 1955-1963 125 & 150cc & GS MODELS (SECOND BOOK OF)
VESPA 1955-1968 GS & SS (BOOK OF)
VESPA 1963-1972 90, 125 & 150cc (THIRD BOOK OF)

VELOCEPRESS MANUALS - MOPEDS & MOTORIZED BICYCLES

CYCLEMOTOR (BOOK OF)
NSU QUICKLY 1953-1963 ALL MODELS (BOOK OF)
PUCH MAXI N & S MAINTENANCE & REPAIR (3 MANUAL COMPILATION)
RALEIGH MOPEDS 1960-1969 (BOOK OF)

VELOCEPRESS MANUALS - THREE WHEELER'S

BOND MINICAR THREE WHEELER 1948-1967 (BOOK OF)
BMW ISETTA FACTORY WORKSHOP MANUAL
BSA THREE WHEELER (BOOK OF)
RELIANT REGAL THREE WHEELER 1952-1973 (BOOK OF)
VINTAGE MORGAN THREE WHEELER (BOOK OF)

VELOCEPRESS MANUALS - AUTOMOBILE BY MAKE

ALFA ROMEO GIULIA WORKSHOP MANUAL 1300 TO 2000cc 1962-1975
ALFA ROMEO GIULIA TECH MANUAL CARBURETED CARS FROM 1962
ALFA ROMEO GIULIA TECH MANUAL FUEL INJECTED CARS FROM 1969
ALFA ROMEO GIULIETTA & GIULIA 750 & 101 SERIES 1955-1965 WSM
AUSTIN-HEALEY SPRITE & MG MIDGET WORKSHOP MANUAL 1958-1971
BMW 600 LIMOUSINE FACTORY WORKSHOP MANUAL
BMW 600 LIMOUSINE OWNERS HAND BOOK & SERVICE MANUAL
BMW 2000 & 2002 1966-1976 WORKSHOP MANUAL
CORVAIR 1960-1969 WORKSHOP MANUAL
CORVETTE V8 1955-1962 WORKSHOP MANUAL
FIAT 500 FACTORY WORKSHOP MANUAL 1957-1973
FIAT 600, 600D & MULTIPLA FACTORY WORKSHOP MANUAL 1955-1969
JAGUAR E-TYPE 3.8 & 4.2 SERIES 1 & 2 WORKSHOP MANUAL
JAGUAR MK 7, 8, 9 & XK120, 140, 150 WORKSHOP MANUAL 1948-1961
METROPOLITAN FACTORY WORKSHOP MANUAL
MGA & MGB OWNERS HANDBOOK & WORKSHOP MANUAL
MG MIDGET TC, TD, TF & TF1500 WORKSHOP MANUAL
PORSCHE 356 1948-1965 WORKSHOP MANUAL
PORSCHE 911 2.0, 2.2, 2.4 LITRE 1964-1973 WORKSHOP MANUAL
PORSCHE 911 2.7, 3.0, 3.2 LITRE 1973-1989 WORKSHOP MANUAL
PORSCHE 912 WORKSHOP MANUAL
TRIUMPH TR2, TR3, TR4 1953-1965 WORKSHOP MANUAL
VOLKSWAGEN TRANSPORTER, TRUCKS & WAGONS 1950-1979 WSM
VOLVO 1944-1968 ALL MODELS WORKSHOP MANUAL

VELOCEPRESS TECHNICAL BOOKS - AUTOMOBILE

FERRARI 250/GT SERVICE AND MAINTENANCE
FERRARI GUIDE TO PERFORMANCE
FERRARI OWNER'S HANDBOOK
FERRARI TUNING TIPS & MAINTENANCE TECHNIQUES
HOW TO BUILD A FIBERGLASS CAR
HOW TO BUILD A RACING CAR
HOW TO RESTORE THE MODEL 'A' FORD
MASERATI OWNER'S HANDBOOK
OBERT'S FIAT GUIDE
PERFORMANCE TUNING THE SUNBEAM TIGER
SOUPING THE VOLKSWAGEN
SOLEX CARBURETORS (EMPHASIS ON UK & EU AUTOMOBILES)
SU CARBURETORS (EMPHASIS ON UK AUTOMOBILES)
WEBER CARBURETORS (EMPHASIS ON ALFA & FIAT)

VELOCEPRESS BOOKS & GUIDES - AUTOMOBILE

ABARTH BUYERS GUIDE
COMPLETE CATALOG OF JAPANESE MOTOR VEHICLES
FERRARI 308 SERIES BUYER'S AND OWNER'S GUIDE
FERRARI BERLINETTA LUSSO
FERRARI BROCHURES AND SALES LITERATURE 1946-1967
FERRARI BROCHURES AND SALES LITERATURE 1968-1989
FERRARI SERIAL NUMBERS PART I - ODD NUMBERS TO 21399
FERRARI SERIAL NUMBERS PART II - EVEN NUMBERS TO 1050
FERRARI SPYDER CALIFORNIA
HENRY'S FABULOUS MODEL "A" FORD
MASERATI BROCHURES AND SALES LITERATURE

VELOCEPRESS BOOKS - RACING

CARRERA PANAMERICANA - MEXICAN ROAD RACE (BOOK OF)
DIALED IN - THE JAN OPPERMAN STORY
IF HEMINGWAY HAD WRITTEN A RACING NOVEL
VEDA ORR'S NEW REVISED HOT ROD PICTORIAL

AUTOBOOKS WORKSHOP MANUALS & BROOKLANDS ROAD TEST PORTFOLIOS

FOR A COMPLETE LISTING OF THE AUTOBOOKS & BROOKLANDS TITLES
THAT WE CURRENTLY HAVE AVAILABLE, PLEASE VISIT OUR WEBSITE.
www.VelocePress.com

Please check our website:

www.VelocePress.com

for a complete up-to-date list of available titles

www.ingramcontent.com/pod-product-compliance
Lightning Source LLC
Chambersburg PA
CBHW070556170426
43201CB00012B/1857